PRAISE FOR *The Restore-Our-Planet Diet*

"A wonderful must-have addition to more than just the cookbook shelf of any conscientious earth dweller. Well researched, user friendly and highly pertinent source like none other. The essential, entirely original in its presentation and **highly recommended,** fact-focused complement to Bittman's *How to Cook Everything Vegetarian*.

This gem of a book fills what has been a hugely important gap. Many are by now well aware of the general idea that—just like in the nutritional realm—environmentally too not all diets are created equal. Yet how to apply this broad understanding to individual meals or dishes has never before been worked out. Scientists like myself found the challenge too applied, and journalists and chefs felt, correctly, that it is outside of their numerical expertise. It took a passionate cook with a PhD in engineering, Dr. Patricia Tallman, to crack this nut. What is so remarkable, and completely novel, about this book is not the useful collection of relatively straightforward, wholesome and delicious recipes, even though the ones I tried are superb. Instead, it is the accompanying detailed comparison of environmental footprint, in terms of averted water use, manure runoff and greenhouse gas emissions, of the vegan and non vegan versions of each recipe. This critical yet elusive actionable information empowers you the reader to make informed decisions that stand to enormously reduce your environmental impact while saving you time, money and the unsettling need to rely on idle speculations or self-serving unsubstantiated assertions. Highly recommended!"
~ Gidon Eshel, Research Professor of Environmental Physics at Bard College, Annandale, NY

"Climate scientists are telling us that even if we were now to put into practice the strictest possible curbs on carbon emissions, the prospects would still be very bleak for hundreds of millions of people. But if we were also to markedly reduce our meat consumption, an entirely different and far more hopeful future would become possible. If we are to create a livable future on this planet, we will need to successfully lessen the demand for meat. And if we do that, it will be because of people like Patricia Tallman, and thanks to those people who read and heed her book."
~ John Robbins, author of *Diet For A New America*; President of The Food Revolution Network

"We're finding more and more evidence that a plant-based diet is the best way to promote a healthy life. And now we are learning that what's best for us is also best for the planet. Eating sustainably means replacing meat and dairy products with delicious, wholesome plant-based foods. The payoff is enormous for your health *and* the Earth. Please read this fascinating and informative guide. Let Patricia Tallman's guidance lead the way, and find the greatness of eating green."
~ Neal Barnard MD, Adjunct Associate Professor at George Washington University's School of Medicine; President of Physicians Committee for Responsible Medicine

"If you care about your health or the environment, then Patricia Tallman's remarkable book, *The Restore-Our-Planet Diet,* is for you. It's the first book I've seen with recipes that include side-by-side comparisons of the nutritional and ecological benefits of plant-based versions over animal-based versions. We learn, for example, that vegan Stuffed Peppers use 466 gallons less water than the animal version, generate 12 pounds less manure, and lower greenhouse gas emissions by the equivalent of driving seven miles. And not only are the benefits compelling, but the recipes are mouth-watering! What better way to change your ecological footprint and improve your health?"
~ David Robinson Simon, author of *Meatonomics.*

"With easy to understand information, graphs and topics, *The Restore-Our-Planet Diet* is a great tool for *anyone* looking to improve their health and the health of the Earth's ecosystems."
~ Keegan Kuhn, Filmmaker, co-director *Cowspiracy*

"A clever new take on what's wrong with eating animals, this book speaks poignantly and practically to the relationship between what's on our plates and the state of our planet, and then tells us how to make changes, simply and deliciously."
~ Victoria Moran, author of *Main Street Vegan* and *The Good Karma Diet*

"Patricia Tallman not only makes a compelling case for replacing animal foods in your diet with choices that protect the planet and your health—she shows you how to do it. And her tips and recipe makeovers are so easy that you'll wonder why you haven't been eating this way all along. This is a wonderful guide for anyone who wants to make choices that protect the future of our planet."
~Virginia Messina, MPH, RD, author of *Vegan for Life, Vegan for Her,* and *Never Too Late to Go Vegan*

The Restore-Our-Planet Diet:

Food Choices, Our Environment, and Our Health

Patricia Tallman, PhD

"The livestock sector emerges as one of the top contributors to the most serious environmental problems, [on] every scale from local to global. The findings of this report suggest that it should be a major policy focus when dealing with problems of land degradation, climate change and air pollution, water shortage and water pollution, and loss of biodiversity . . . The impact is so significant that it needs to be addressed with urgency."

- "Livestock's Long Shadow," Food and Agriculture Organization of the United Nations, 2006

Print ISBN-10: 1508487626

Print ISBN-13: 978-1508487623

Publishing assistance provided by LOTONtech (www.lotontech.com)

To Ralph, my beloved, who transformed my life.

This book is written in the fervent hope that each animal may live the life to which every individual is entitled, free of human exploitation.

"Besides agreeing with the aims of vegetarianism for aesthetic and moral reasons, it is my view that a vegetarian manner of living by its purely physical effect on the human temperament would most beneficially influence the lot of mankind." ~ Albert Einstein

TABLE OF CONTENTS

ACKNOWLEDGEMENTS IX

INTRODUCTION 1

CHAPTER 1
The Environmental Costs of Meat and Dairy Consumption........ 7

CHAPTER 2
The Health Impacts of Meat and Dairy Consumption........ 19

CHAPTER 3
How Veganizing Beef Recipes Improves Environmental and Nutritional Parameters: *An Analysis*........ 33

CHAPTER 4
How Veganizing Pork Recipes Improves Environmental and Nutritional Parameters: *An Analysis*........ 51

CHAPTER 5
How Veganizing Chicken Recipes Improves Environmental and Nutritional Parameters: *An Analysis*........ 63

CHAPTER 6
How Veganizing Dairy and Egg Recipes Improves Environmental and Nutritional Parameters: *An Analysis*........ 77

CHAPTER 7
What Are Humans Designed To Eat?........ 95

CHAPTER 8
Vegetables........ 101

CHAPTER 9
Pulses and Grains........ 121

CHAPTER 10
Soy (Tofu)........ 139

CHAPTER 11
Baking Without Dairy or Eggs........ 153

RESOURCES........ 169

APPENDIX........ 173

REFERENCES........ 181

ACKNOWLEDGEMENTS

I would like to thank Registered Dietitian Matt Ruscigno, MPH, RD, for contributing Chapter 2 of this book. A former Chair of the Vegetarian Practice Group of the US Academy of Nutrition and Dietetics, Matt is an endurance athlete and a vegan. I also wish to thank Registered Dietitian Mark Rifkin, MS, RD, for calculating the Nutritional Benefits Charts and for writing the Nutritional Benefits Summaries in Chapters 3 through 6.

I am very grateful to my editor Lisa Ferdman, desktop publisher Richelle Benoit, publishing consultant Tony Loton, cartoonist Graham Harrop, and website designer Cam Dore, for their valuable contributions.

Particular gratitude goes to Jim Tallman, Taruna Goel, and Vesanto Melina, for their guidance and encouragement.

INTRODUCTION

> *"A vegetarian in a Hummer produces fewer greenhouse emissions than a meat eater in a Toyota Prius."*
>
> - *Meat the Truth* (documentary), 2008

The environmental problems first publicized fifty years ago in Rachel Carson's *Silent Spring* now have reached a critical stage. Former US Vice President Al Gore's compelling book, *An Inconvenient Truth* (2006), and his award-winning documentary film of the same title have contributed significantly to the ever-growing public awareness of the dangers of greenhouse gases. The Intergovernmental Panel on Climate Change's latest Synthesis Report (November 2014) indicates, with 95 percent certainty, that nearly all global warming occurring since the 1950s has resulted from human activity. The most readily apparent consequences include rising sea levels, ocean warming and acidification, increased global temperatures, and melting Arctic Sea ice. Additionally, geoscientists assert that climate change has contributed to the larger, more long-lasting wildfires witnessed in recent years. In May 2014, a US National Climate Assessment report also showed a rise in the number of wildfires across the Western parts of the United States.

As conscientious citizens, many of us seek solutions from our governments regarding industrial practices and environmental protection but are disappointed by officialdom's seeming preoccupation with the economic bottom line. Yet, one of the best solutions to a large array of ecological concerns lies well within our reach. Indeed, it is as close as our dinner plates. The conventional Western diet of meat and dairy products has depleted our planet's natural resources, thus betraying our own best interests. Our virtual dependence upon the livestock industry is among the chief causes of water pollution, global warming, land degradation, habitat loss, wildlife extinction, rainforest destruction, antibiotic resistant bacteria, and food availability. This book provides a unique understanding of our current predicament, and offers practical tools to effect measurable change.

Despite having earned a Doctorate in Water Resources Engineering and a Master's degree in Environmental Sciences, I was, at first, unaware of the impact of my daily meat and dairy consumption. It had not occurred to me that what I ate profoundly affected my health, the environment, and animal welfare. I had devoted myself to solving micro-environmental problems, but unwittingly was contributing to nature's degradation. Perhaps other eco-conscious people may be equally uninformed regarding the damage wrought by the livestock industry.

The resources consumed by this industry truly are staggering. For instance, 100 times more water is required to produce animal protein than to produce plant protein. A meat-based diet uses seven times more land than does a plant-based diet. The energy needed for meat production generally is 10 times greater than that for vegetable, grain, and legume production.

A study commissioned by the Vegetarian Resource Group states that approximately five percent of the current American population is vegetarian, and that half of the people in this group are vegans. Since the US population is 320 million, 16 million Americans do not eat animals. This trend is expected to grow.

If everyone on Earth were to follow the Western diet, it would require four planet Earths to support the current population of 7 billion. How will we secure food production for future generations when our population reaches 9 billion, as is projected for the year 2050? Clearly, if we continue to degrade the biosphere to the point that agriculture is poorly supported, we will be unable to feed ourselves. The research, insights, and recommendations to be found in these pages will prove invaluable to thinking consumers.

If everyone on Earth were to follow the Western diet, it would require four planet Earths to support the current population of 7 billion.

The beginning chapters present an overview of the environmental and health consequences of our prodigious meat and dairy consumption, and provide a convincing case for a whole-food, plant-based diet. Subsequent chapters compare traditional "comfort food" recipes to their vegan counterparts. The accompanying analyses demonstrate that we can reduce our eco-footprint (i.e., water usage, manure production, and greenhouse gas emissions) while simultaneously *increasing* the nutritional benefits of our diet (considering total calories, protein, total fat, saturated fat, cholesterol, iron, and sodium).

Next, an examination of human physiology assists readers in determining whether our anatomical design is more suited to eating animal flesh or plants.

The tempting plant-based recipes offered, featuring vegetables, grains, and legumes, are intended to address the epidemic of chronic diseases in Western populations that is directly related to food choices. Obesity, heart disease, stroke, cancer, diabetes, chronic kidney failure, and other conditions exact a huge toll—when, in fact, some of them are preventable or reversible. Our lifestyle choices have cost the collective Canadian and American economy at least $296 billion in annual healthcare expenses.

The adoption of a whole-food, plant-based diet, if undertaken by sufficient numbers of people, will decrease environmental damage and healthcare costs significantly. There is no reason to wait for governments to implement policy changes before making our own powerful decisions; in truth, there is every reason *not* to wait. This book will enable people who care deeply about our world to take practical steps to restore it—and themselves—to optimal health.

Our lifestyle choices have cost the collective Canadian and American economy at least $296 billion in annual healthcare expenses.

A comprehensive discussion of seafood necessarily would encompass such issues as destructive fishing techniques, the spreading of illnesses from fish farms to wild stocks, overfishing, marine pollution, ocean acidification and dead zones, ocean warming, and rising sea levels. While there also are health reasons for leaving seafood off the menu—including heavy-metal contamination and high cholesterol—this book focuses only upon the consequences of land-based animal agriculture. There now are vegan alternatives to seafood whose tastes and textures are very similar to those of fish, shrimp, scallops, and crab.

When adopting a plant-based diet, the first step is to recognize that it's not an "all or nothing" regime. While some people are able to switch fairly quickly to a completely plant-based diet, others may need to proceed more slowly, over a period of months. Still, others may retain a minor portion of animal protein in their diet. The important point is that any replacement of animal protein with plant protein is better than none. The more plant-based the diet, the greater the benefits to the environment, the animal population, and our health.

The important point is that any replacement of animal protein with plant protein is better than none. The more plant-based the diet, the greater the benefits to the environment, the animal population, and our health.

If you cannot transition to a completely plant-based diet in six to twelve months, it is far better to replace half of your animal foods with plant foods in that time than not to undertake any change at all. Don't give up the opportunity to do something simply because you cannot do *everything*.

The advantages of adopting an eco-healthful diet include improving your physical health, energy, and mental clarity; expanding your culinary tastes; reducing the escalating national healthcare costs; decreasing air and water pollution from factory-farm runoff; freeing up large amounts of grain, land, energy, and water resources to grow food for the world's hungry and for future generations; and saving the 70 billion land animals worldwide that are slaughtered for food each year. Such transformation is entirely possible.

How light will *your* eco-footprint be?

SECTION I

Why a Plant-Based Diet?

"Talk about stinky emissions!"

Chapter I

The Environmental Costs of Meat and Dairy Consumption

> *"Just moving away from meat for one day a week is more effective [in terms of energy efficiency and environmental health] than buying everything you eat locally."*
>
> *Harvard Business Review, 2011*

Livestock agriculture imposes a devastating toll upon our environment in many ways: greater water and energy consumption; water and air pollution; deforestation; habitat destruction; wildlife species extinction; and climate change. All of these contribute to the degradation of our planet and harm all who live in it – humans and animals alike.

GREENHOUSE GASES AND CLIMATE CHANGE

Climate change is on many people's minds. Global organizations such as the United Nations, National Audubon Society, Worldwatch Institute, Sierra Club and the Union of Concerned Scientists all suggest that animal agriculture is more damaging to the environment than any other human activity. In fact, the UN Environment Program stated in 2010 that a global shift toward a vegan diet is *critical* in order to mitigate the worst impacts of climate change. The UN's Food and Agriculture Organization's (FAO) 2006 Report entitled *Livestock's Long Shadow* revealed that 18 percent of global warming was attributable to animal agriculture. This figure is greater than that for all modes of transportation combined. A more recent estimate shows that the true figure may be as high as 25 percent when factors such as the burning of tropical rainforests and pollution are taken into account.

The three main greenhouse gases (GHGs) are carbon dioxide (CO_2), methane, and nitrous oxide; their combined emissions are measured in CO_2 equivalents, or CO_{2e}. Animal agriculture accounts for 9 percent of global CO_2 emissions, as well as for 40 percent of global methane and 65 percent of global nitrous oxide. The latter two are more potent GHGs. The higher percentages of these more potent GHGs—methane at 23 times and nitrous oxide at 300 times that of CO_2—highlight the dire consequences of the livestock industry's contribution. In

December 2013, BBC News reported that cattle are the greatest source of GHGs, accounting for over 75 percent of global livestock emissions.

The Environmental Working Group's data (Figure 1) on full lifecycle GHG emissions for common animal and plant foods, show that animal foods emit much greater GHGs (in CO_{2e}) than do plant foods–in some cases, 10 times as much. ("Lifecycle" refers to the cradle-to-grave GHG emissions engendered by the fertilizers and pesticides used to grow feed, the animals' factory farms, transportation, slaughterhouses, meat processing, final consumption and unused food disposal.)

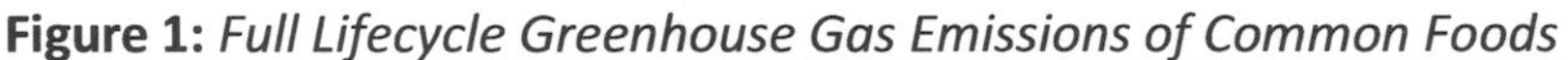

Figure 1: *Full Lifecycle Greenhouse Gas Emissions of Common Foods*

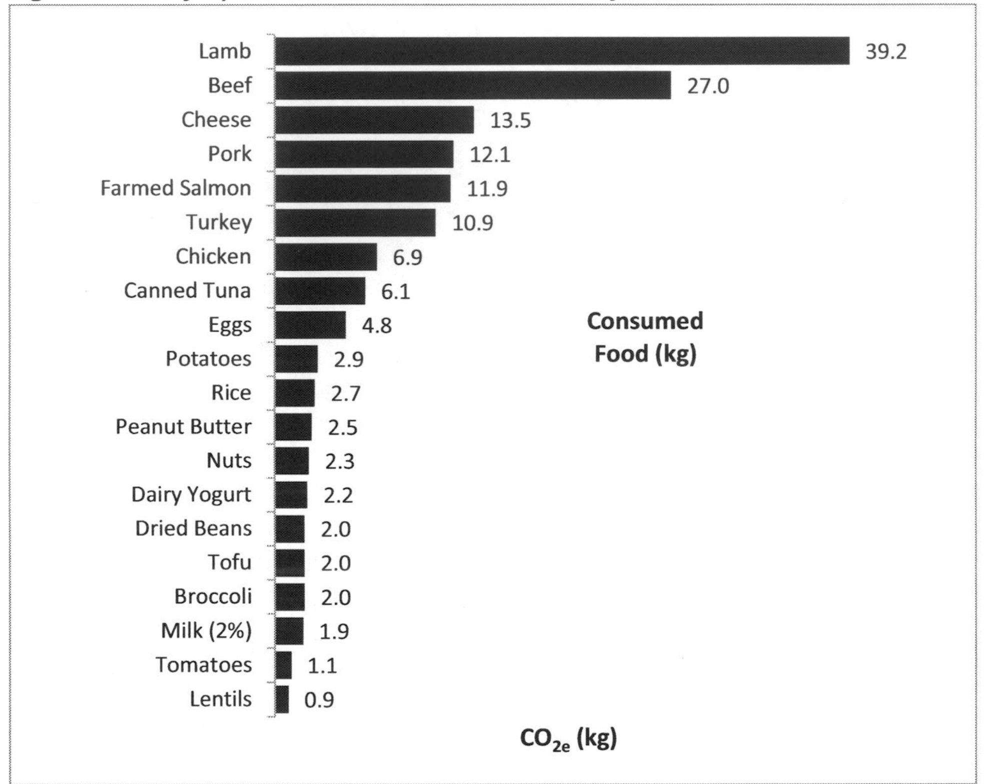

Data from Environmental Working Group, (Meat Eater's Guide: Report 2011)
Post Farmgate Emissions (includes processing, transport, retail, cooking, and waste disposal)

Lamb and beef production top the list as the highest emitters of greenhouse gases (39.2 and 27 kg CO_{2e}, respectively, emitted per kg of food), followed by cheese, pork, farmed salmon, and turkey. Other animal food production, such as chicken, tuna, and eggs, results in lower emissions, but even the lowest of these meat products (eggs, at 4.8) is higher than those for plant foods, all of which are below 3 kg CO_{2e}. This means that we can achieve meaningful GHG savings by replacing animal foods, particularly lamb, beef, and cheese, with plant alternatives. The reason for the particularly high GHG emissions for lamb, beef, and cheese is that the stomachs of ruminant (i.e., cud-chewing) animals are essentially chambers for methane production.

"Humane" Meat and Greenhouse Gas

An ever-increasing number of conscientious consumers are striving to minimize environmental impact by choosing so-called humane meat over conventional factory-farmed products. "Humane meat" usually comes from smaller, and often organic, farms. Although the consumers' choice is well-intentioned, "humane" meat hardly represents a true solution when one considers the environmental perspective. The more time it takes for a pasture-raised animal to attain the desired weight for slaughter, the more opportunity it has to produce GHGs. According to Professor Gidon Eshel of Bard College in New York, grazing animals emit *two to four times more* methane than their factory-farmed counterparts. And since methane is 23 times more potent than CO_2, it's easy to see why "humane" meat is environmentally more costly. In addition, greater land resources are required for pasture-raised animals. Consequently, an attempt to apply this model on a larger scale, to accommodate the current demand for animal flesh, would be highly impractical.

Eating Local Products

A recent, popular trend promotes the purchase of locally-sourced foods. Unfortunately, the GHG savings generated by this laudable approach is only meaningful for the transport of non-animal products. A 2008 study by researchers Christopher Weber and Scott Matthews at Carnegie Mellon University found that most (83 percent) of the GHG emissions for animal products occurred during the *production phase*: from growing the crops and raising the animals. Transportation (of grain, feed, and supplies, and of animals to the slaughterhouse) accounted for only 11 percent of the emissions – a mere 4 percent of which represented the final delivery of the product. The remaining 5 – 6 percent was attributed to wholesaling and retailing. Consumers and local-food advocates who speak of transportation's damaging effects usually consider only the product's final delivery (i.e., from production facility to retail store), rather than the total distances travelled. Thus, a more effective strategy to reduce GHGs would be a dietary shift from animal products to plant proteins, rather than buying locally raised meat.

Transportation (of grain, feed, and supplies, and of animals to the slaughterhouse) accounted for only 11 percent of the emissions – a mere 4 percent of which represented the final delivery of the product.

The same study also found that within food production, over half of the GHG emissions were non-CO_2, namely methane (23%) and nitrous oxide (32%)—which, as previously stated, are 23 times and 300 times respectively, more potent than CO_2. And the major contributors to methane emissions are ruminant animals such as cows, sheep, and goats (i.e., red meat and dairy production). Nitrous oxide emissions are caused by fertilizer applications and by manure management. Thus, while buying local would achieve a few percent reduction in GHGs, an easier, more practical, and healthier way to reduce our carbon footprint would be to replace an animal-based diet (particularly red meat and/or dairy) with a plant-based one. The latter is an environmentally responsible choice.

Comparison of GHGs for Different Diets

A 2013 study from the UK, published in *Climate Change* journal, found that participants in the EPIC-Oxford cohort study who were regular meat-eaters contributed 46 to 51 percent more food-derived GHGs than fish-eaters did, and 50 to 54 percent more than vegetarians did. The biggest difference was that between meat-eaters and vegans: meat-eaters contributed nearly 100 percent more food-related GHGs than did vegans. In fact, high meat-eaters contributed about 7.2 kg of CO_{2e} per day, while vegetarians contributed 3.8 kg CO_{2e} and vegans, 2.9 kg CO_{2e}.

WATER RESOURCES

But even graver than carbon emissions are the dwindling global freshwater resources we are facing. Currently, animal agriculture uses about 50 percent of the world's freshwater resources. This percentage is higher than the 18 – 25 percent GHG emissions by the livestock industry. Increasing competition among industrial, agricultural, and urban demands for water contributes to this growing crisis. The instability of hydrological patterns that is one of the effects of climate change serves to exacerbate the problem.

Water now is regarded as being as precious a commodity as oil. The UN FAO predicts that, by 2025, two-thirds of the world's population will live in water-stressed areas (defined as having less than 1700 cubic meters per capita per year). As indicated in the Introduction, if our global population of 7 billion were all to eat a meat-based diet, four planet Earths would be required to meet this demand. Clearly we don't have that option. Similarly, if we don't drastically reduce or eliminate our consumption of animal products now, how will the projected 9 billion people in 2050 have enough food to eat?

Water scientists at the Stockholm International Water Institute have suggested that, given the demands that animal agriculture imposes upon the world's water supply, we should limit our consumption of animal-based foods to just 5 percent of our total caloric intake. This would ensure that there is enough water to grow food. Such a serious warning should impel both policy makers and consumers who care about our planet to transition to a plant-based diet.

Water Requirements for Various Diets

An examination of three types of diets will allow us to compare their usage of water. Producing one day's worth of food for a person on a plant-based diet requires roughly 300 gallons of water. For a lacto-ovo vegetarian, that requirement jumps to 1,200 gallons; whereas, for an omnivore, the figure skyrockets to 4,000 gallons. And this is just for one person for one day of the year. If everyone were to switch to a plant-based diet, the aggregate water savings would be enormous.

Water Footprint

Figure 2 shows the water footprint for common animal and plant products. It is evident that water footprints for vegetables are significantly lower than those for animal flesh, especially when compared to those of ruminant animals.

Figure 2: *Water Footprint of Animal and Plant Foods*

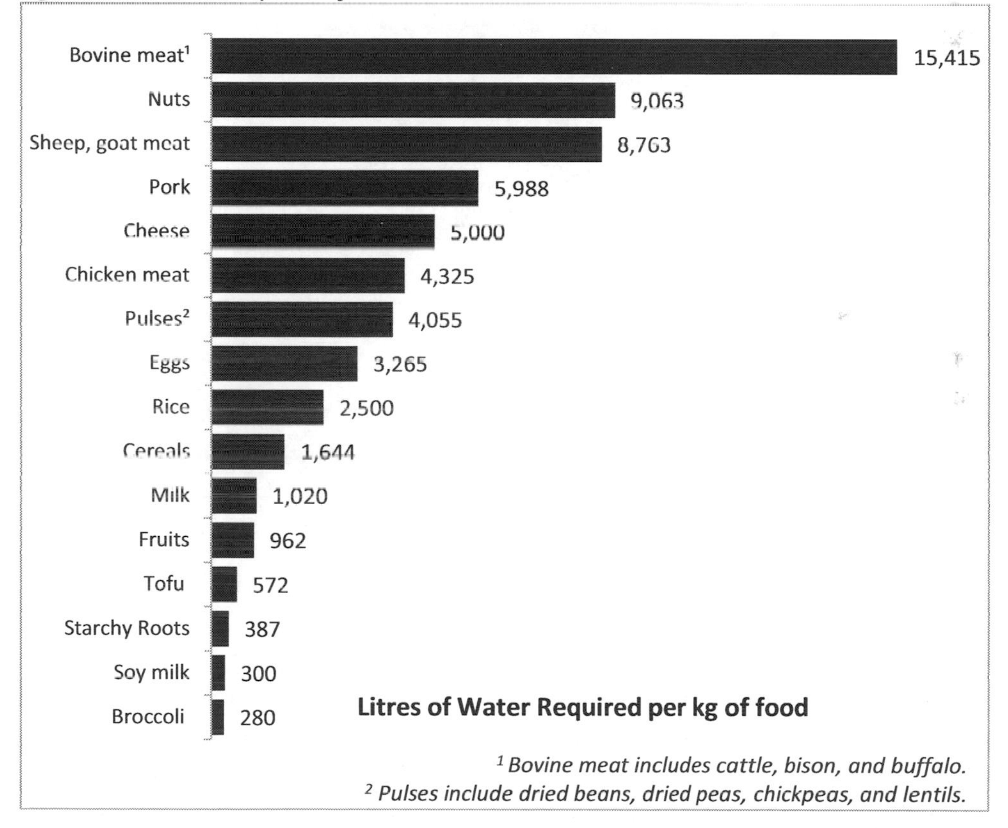

Sources: Data from Arjen Y. Hoekstra, Twente Water Center, University of Twente, "The hidden water resource use behind meat and dairy", Animal Frontiers, April 2012, Vol. 2, No. 2 – Table 1; M.M. Mekonnen and A.Y. Hoekstra, "The green, blue and grey water footprint of crops and derived crop products" Hydrology and Earth System Sciences, 15, 1577-1600, 2011; Table 2 in Appendix.

The disproportionately high water footprint for ruminant animal products (beef, lamb, sheep, and goat) stands out. Although nuts also have a high water footprint, they are not generally consumed in large quantities. According to David Simon, in his book *Meatonomic$*, even the 10 percent water savings attained by raising cattle organically uses an amount of water that is many times greater than that used to produce plant foods.

While it may appear, from first glance at Figure 2, that the water footprint for chicken is similar to that for beans, other factors must be considered. Data from Dutch researchers Mekonnen and Hoekstra indicate that the amount of protein per kg of food that is yielded by plant sources is higher than that from animal sources. Figure 3 shows that pulses yield 215 grams of protein per kg, while beef yields 138 grams of protein per kg. Chicken meat yields even less at 127 grams, which is only 60 percent of that for beans.

Figure 3: *Protein Yield of Various Foods*

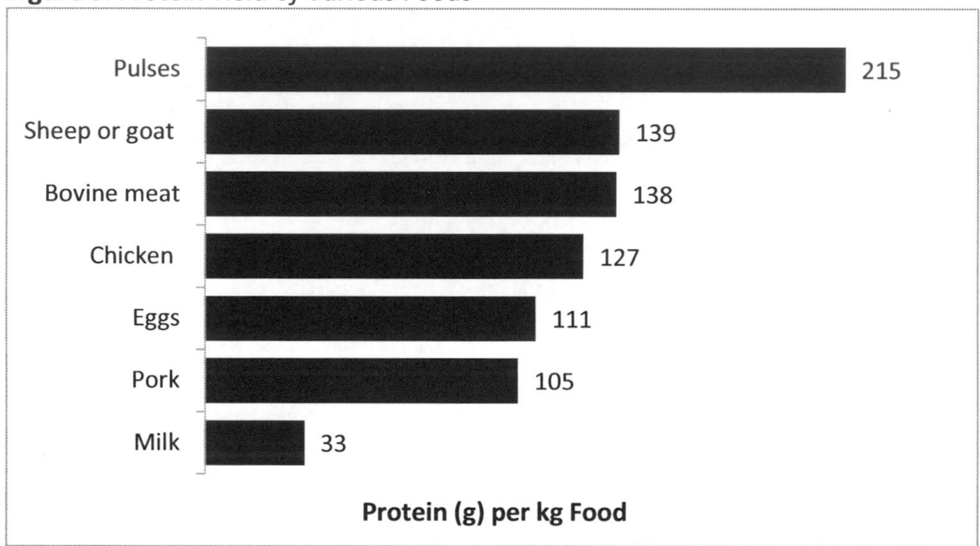

Source: Data from Arjen Y. Hoekstra, Twente Water Center, University of Twente, "The hidden water resource use behind meat and dairy", Animal Frontiers, April 2012, Vol. 2, No. 2 – Table 1

Water Footprint in Relation to Protein

Figure 4 further illustrates the lower water footprint of plant foods when measured with respect to their protein content. If we compare chicken to beans, we note that chicken has a higher water footprint (34) per gram of protein than beans (19) per gram of protein. This is in spite of the fact that chicken's water footprint of 4325 liters is similar to that of beans at 4055 liters. This infers that we can choose healthier plant proteins that require less water usage and emit less GHGs. (For an explanation of the "incomplete" protein myth, see Chapter 2.)

Figure 4: *Water Footprint with respect to Protein Content*

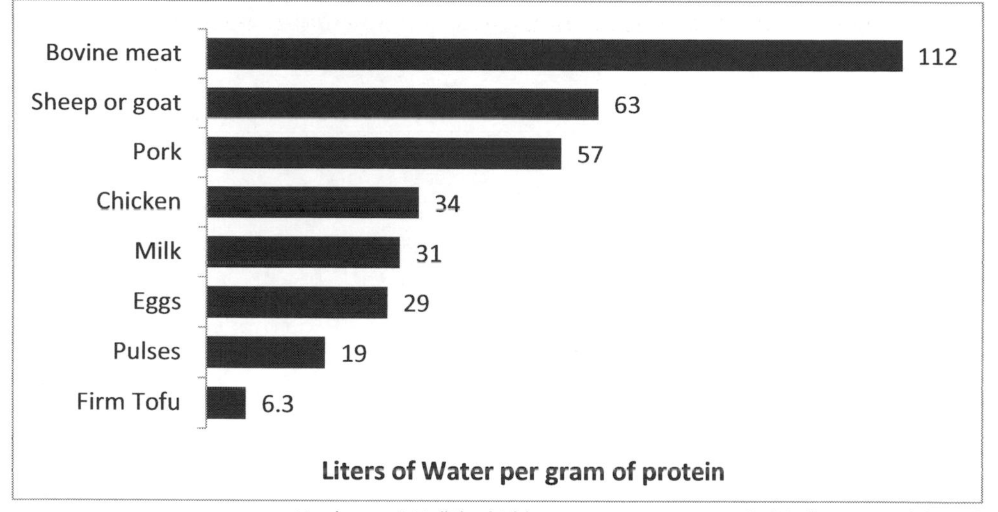

Hoekstra, A.Y. "The hidden water resource use behind meat and dairy", Animal Frontiers, April 2012, vol. 2, no. 2, Table 1.

ANIMALS' INEFFICIENT CONVERSION OF PLANT PROTEIN

As with humans, animals' necessary metabolic processes result in an inefficiency when converting plant protein into flesh. According to Vaclav Smil, Distinguished Emeritus Professor from the University of Manitoba, it takes 3.3 kg of feed (mostly corn and soy) to produce 1 kg of chicken meat. This feed amount rises to 9.4 kg for 1 kg of pork, and to an astronomical 25 kg of feed for 1 kg of beef. Of course, these amounts vary somewhat according to crop types, climatic conditions, geographical differences, and differences in animal breeds. Nonetheless, the relative inefficiency is difficult to dispute. Imagine the many hungry people we could feed directly with these grains if we did not feed the grains to animals, which are then slaughtered as food.

FOOD AVAILABILITY

The global cattle population of roughly 1.3 billion consumes an amount of calories (chiefly in grains) that could feed 8.7 billion people! Such an amount not only would provide food for today's poor and hungry population, but also would represent food security for the future. Emily Cassidy, lead researcher at the University of Minnesota at St. Paul, indicated that globally, shifting crops away from animal feed and biofuel to direct human consumption could feed an additional 4 billion people. Currently, there are 1 billion hungry people worldwide; 6 million children die every year from malnutrition. At the same time, 2.1 billion people in the world are overweight or obese, primarily as a consequence of the Western diet of meat and dairy. That means we have twice as many people overfed than underfed. Even if the contributing causes of starvation (food distribution, politics, and world economies) were to be resolved, a continued reliance on meat and dairy products would squander the critical food resources needed to feed people in developing nations. Thus, a global shift toward a plant-based diet is an essential component of any strategy to mitigate or to relieve world hunger.

ANIMAL WASTE

Today, humans kill about 70 billion farmed animals worldwide for food each year. Where does all the runoff from industrial animal agriculture (i.e., factory farms) go? This runoff contains not only manure, but also pesticides, chemical fertilizers, antibiotics, and hormones. Where the manure is not kept for use as fertilizer, it pollutes waterways and airspace. In the US, farm animals produce 130 times as much waste as does the nation's human population. Runoff from factory farms pollutes more waterways than all other industrial sources combined. In 1995, when a lagoon at a North Carolina hog farm that contained eight acres of hog manure burst, it spilled 25 million gallons of hog urine and feces into the New River. As a result, between 10 and 14 million fish died. In Iowa alone, hog factories produce over 50 million tons of waste annually. While some manure lagoons are lined in order to prevent groundwater contamination, much of the manure is sprayed onto fields, inevitably polluting streams and groundwater. Professor Gidon Eshel noted that if producers were required to pay to mitigate the costs of pollution, this expense undoubtedly would be passed on to consumers. Any such policy would likely change the entire food production model and consumer behavior, as well. But perhaps this is what is needed.

Figure 5: *Manure Produced During Manufacturing of Animal and Plant Foods*

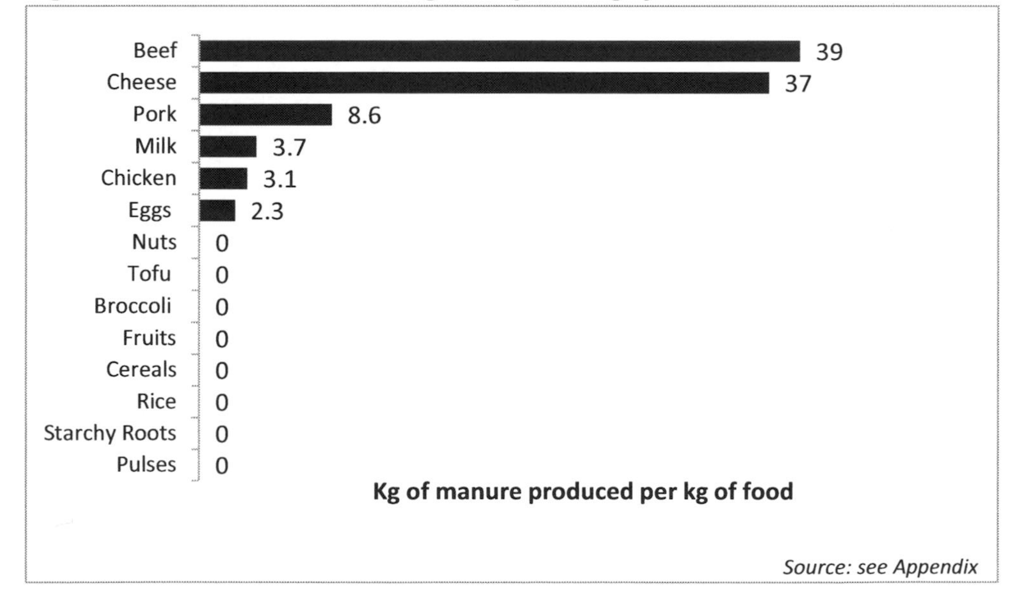

Nutrients from Animal Waste

The nitrogen and phosphorus from animal waste pollute waterways, resulting in algae blooms. When algae die, they sink to the ocean bottom. The consequent depletion of the oxygen supply ultimately kills sea life. There are now over 400 dead zones around the globe. The Gulf of Mexico's dead zone covers 21,000 square kilometers. In addition, animal waste pollutes the air with hydrogen sulphide and ammonia gases.

LAND USE

Globally, 70 percent of agricultural land is dedicated to the feeding and raising of livestock; in the US that number is higher still, at 80 percent. A meat-based diet requires, on average, seven times more land than does a plant-based diet. Furthermore, the UN Food and Agriculture Organization predicts that, by 2030, the demand for meat and dairy will increase by 68 and 57 percent, respectively. Yet, if agricultural land were used to grow food for direct human consumption rather than for the cultivation of feed crops for animals and grazing, there would be an enormous increase in national and global food supplies.

RAINFOREST DESTRUCTION

Tropical rainforests contain 80 percent of the world's plant species and 50 percent of its animal species. The true lungs of our planet, these incalculably rich regions once occupied 14 percent of our land mass but have now declined to an alarming 6 percent. Eighty percent of this destruction is a result of cattle ranching to satisfy the demand for beef and of monocrops, such as soy, grown to feed these animals.

Twenty percent of the Amazon rainforest existing before European colonization has been destroyed. At a disappearance rate of 20,000 square miles per year, it is predicted that it will be gone within 50 years. David Kaimowitz of the Centre for International Forestry Research observed, "In a nutshell, cattle ranchers are making mincemeat out of Brazil's Amazon rainforests." In Central America, 40 percent of the rainforests have been destroyed, again mostly for the purpose of cattle grazing.

Is the burger worth all this destruction? What value do we place on the rainforests? If recognition of the rainforest's value is factored into the calculation of the cost of a burger, it could hardly be regarded as cheap fast food. Instead, it would become an exorbitantly priced commodity, out of reach for all but a very wealthy few.

Loss of Biodiversity

The rainforest's biodiversity may be illustrated by the fact that a naturalist once found 700 species of butterflies within a 3-mile radius in the Amazon. By comparison, there are approximately 321 butterfly species on the entire European continent. Rainforest plants are studied for their medicinal properties; a quarter of the world's medicine comes from these plants. And of the plants with anti-cancer properties, 70 percent of them exist only in the tropical rainforests.

ANTIBIOTIC USE

Eighty percent of antibiotics sold in the US are destined for the livestock industry, chiefly for healthy animals. The routine and irresponsible use of these drugs to promote the growth of factory-farmed animals in unsanitary conditions and to prevent or to curb disease has led to the proliferation of superbugs that are resistant to antibiotics prescribed for humans. According to *The New York Times,* a 2005 study found that the antibiotic-resistant bug MRSA kills almost 19,000 people in the US annually. Clearly, the excessive use of antibiotics in animals has had the unintended effect of reducing their efficacy in humans when prescribed to combat infectious diseases such as pneumonia. The American Medical Association, the American Academy of Pediatrics, the American Society for Microbiology, and the American Public Health Association all have expressed their objections to the widespread, non-therapeutic administration of antibiotics to farmed animals, so as to protect public health.

According to The New York Times, a 2005 study found that the antibiotic-resistant bug MRSA kills almost 19,000 people in the US annually.

Rise of Superbugs in Human Infectious Diseases

Recently, convincing evidence from Canada has resulted in a ban (effective May 15, 2014) on the use of ceftiofur (an antibiotic) injections into eggs at hatcheries. As early as 2003, microbe trackers from Canada's Public Health Agency reported higher rates of ceftiofur resistance in Quebec. One year later, the same resistance was noted in Ontario, in both humans and chickens. In 2010, *The Journal of Emerging Infectious Diseases* reported a strong correlation between resistant *Salmonella Heidelberg* (a strain of bacteria that resists ceftiofur) in chicken meat at Canadian retail stores and the incidence of ceftiofur-resistant bacteria in humans across the nation. In that same year, the FDA found that half of the chicken breasts tested had antibiotic-resistant *E. coli.* This compelling research demonstrates that antibiotic use in livestock leads to resistant microbes in people. The evidence was used to implement European and US bans on agricultural use of ceftiofur.

NITROGEN EMISSIONS

Half of today's food is grown with synthetic, nitrogenous compounds, and this dependence on chemicals is increasing. Nitrogen fertilizers enter the environment in the form of nitrous oxide (300 times more potent than CO_2), via water contamination and atmospheric GHGs. Professor Emeritus Vaclav Smil of the University of Manitoba indicated that the release of nitrogen is expected to increase, but that our dependence on chemicals can be reduced *if* we reduce our animal food intake. Unfortunately, as the standard of living increases because of urbanization and industrialization, traditional diets of cereals, tubers, and legumes typically are replaced by meat-centred diets. This change results in increased nitrogen emissions into the environment—precisely the opposite of the desired trend.

WILDLIFE EXTINCTION

What is the relationship between livestock agriculture and wildlife extinction? When wilderness—ecosystems such as forest, wetlands, marshes, or grasslands—are destroyed in favor of animal agriculture, wild animals become displaced. Habitat loss resulting from deforestation and grassland destruction is the primary cause of species extinction. For example, mixed-tree habitats are clear-cut in order to grow monoculture grass for cattle grazing. In central Canada, much of the prairies are devoted to agriculture. The black-footed ferret and the prairie swift fox have become extinct in Canada, and 14 other prairie species are either endangered or threatened. According to the Natural Resources Defense Council, the US Department of Agriculture's Wildlife Services Division (USDA WSD) spends over $100 million annually to kill some 100,000 carnivores as part of the "predator control" program to appease agricultural interests. These include lethal methods, such as trapping, aerial gunning, poisoning, and denning (killing young in their dens) as well as non-lethal methods. Yet the USDA WSD's records attribute most livestock losses to reasons other than predation: namely weather, disease, illness, and birthing complications.

PESTICIDE USE

Pesticide use on feed crops is a familiar environmental concern. Seventy percent of grain grown in the US is intended for livestock consumption. The production of genetically-modified (GM) food crops in the US, such as corn and soy, has led to increased herbicide use. This practice flies in the face of industry claims of improved yields, reduced costs, and lesser environmental impacts. Critics have warned that widespread planting of GM crops with resistance to herbicides containing glyphosate is likely to result in superweeds that are similarly resistant. Glyphosate is the killing agent present in these herbicides: Monsanto's Roundup,

Bayer's Garden, Syngenta's Touchdown, and Dow's Durango. Marian Nestle wrote in *Safe Food: The Politics of Food Safety* that weeds resistant to Monsanto's Roundup started to appear in Georgia in 2004, and quickly spread to other southern states. By 2009, over 100,000 acres in Georgia were infested with Roundup-resistant pigweed. Nestle noted that, ironically, farmers were advised to apply multiple herbicides in an effort to control the problem, thus defeating the original goal of minimizing herbicide applications. In the US, weed resistance now has spread to 14 million acres.

SUMMARY

The environmental costs of meat and dairy consumption are many and far-reaching. They include greenhouse gas emissions that precipitate climate change, the egregious overuse of water, the commitment of vast land and energy resources, the perpetuation of world hunger through irresponsible use of resources, water and air pollution, deforestation, biodiversity loss, wildlife extinction, rise of superbugs, and the creation of superweeds. The consequences are devastating to our shared environment – and some may well be irreversible.

What about the health implications of the Standard American Diet centered on meat and dairy? This will be discussed in the next chapter.

Chapter 2

By Matthew Ruscigno, MPH, RD

The Health Impacts of Meat and Dairy Consumption

The beef industry has contributed to more American deaths than [have] all the wars of this century, all natural disasters, and all automobile accidents combined. If beef is your idea of 'real food for real people,' you'd better live real close to a real good hospital."

Neal D. Barnard, M.D., President, Physicians Committee for Responsible Medicine, Washington, D.C.

The concept of eating plant foods to treat or prevent disease is not new. Long before nutrients were discovered and before nutrition existed as a field of science, "plant matter" diets were prescribed for a host of maladies. As medicine progressed, the importance of adequate nutrition became recognized, and vitamin and mineral deficiencies became subjects of study.

Although undernourishment is far less common in North America today, millions of people die of preventable chronic diseases related to diet and lifestyle. Alarmingly, these death rates are increasing. Nutritional science now has identified the beneficial components of each food, the relationship between diet and disease prevention, and the means of effecting appropriate changes. Nonetheless, healthcare providers and the public alike remain largely unaware that increased consumption of plant foods improves long-term health.

The Diet and Disease Connection

Nearly half of all deaths in the US are caused by heart disease or cancer. These two leading causes, in that order, are followed by stroke, Alzheimer's disease, diabetes mellitus, and hypertension. Thus, nearly two thirds of US deaths stem directly from avoidable risk factors: namely, improper diet and inadequate exercise. Canadian statistics are equally concerning; cancer and heart disease together account for 50 percent of all deaths, with cancer as the number one cause. Furthermore, diabetes and Alzheimer's disease also rank among the top 10 causes. This is not surprising when one considers that the Standard American Diet—whose acronym, appropriately, is SAD—is composed of refined grains, refined sugar, refined oils, and fats, and is centered on meat and dairy products.

While these food choices provide sustenance, they do not promote health. The SAD diet is associated with many of the chronic diseases most prevalent today.

Not only do chronic diseases claim millions of lives, but they cost tens of billions of dollars in healthcare. A 1995 article in *Preventive Medicine* estimated that the medical cost associated with meat consumption is between 28 and 61 billion dollars per year. A 2007 study in *Obesity* estimated the economic impact of the cardiometabolic risk factors—diabetes, hypertension, hyperlipidemia, and obesity—together at being over 80 billion US dollars a year.

The SAD diet is associated with many of the chronic diseases most prevalent today.

How Diet Affects Health

The largest studies on diet and health track the diets of large populations over extended periods and then measure the health outcomes. Researchers seek answers to the following questions: do people who die of heart attacks have common factors in their diets? How do their diets differ from those of individuals who live to be 90 years old? It turns out that there are, indeed, dietary components that are associated with lower rates of disease—almost all of which are found in plant foods. The Adventist Health Study 2, an ongoing research project involving nearly 100,000 people, found that vegetarians and vegans live longer, on average, than their meat-eating counterparts and have statistically lower rates of all cardiovascular diseases. Other research similarly suggests that plant-based diets are helpful in preventing and, in some cases, reversing disease.

Cancer

Cancer is a complex disease whose potential risk factors include lifestyle, environmental, and genetic components. The consensus among experts is that good nutrition and adequate physical activity play key roles in the prevention of most cancers, especially prostate and colorectal cancers. A 2008 paper entitled, "Cancer is a preventable disease that requires major lifestyle changes" argues that poor nutrition is responsible for more cases of cancer than tobacco. Its authors recommend increasing consumption of fruit, vegetable, and whole grain, as well as reducing intake of red meat and fried foods. The World Cancer Research Fund International, a leading authority,

A 2008 paper entitled, "Cancer is a preventable disease that requires major lifestyle changes" argues that poor nutrition is responsible for more cases of cancer than tobacco.

advises that the best measures for cancer prevention are maintaining a healthy weight, getting enough exercise, limiting processed energy-dense foods and red meats, avoiding processed meats, and increasing consumption of vegetables, fruits, whole grains, and beans. A study in the *Journal of Clinical Oncology* indicates that, simply by adding more fruits and vegetables to the diet of the American populace, 20 percent of cancer cases could be prevented and 200,000 lives could be saved each year.

Cardiovascular Disease (CVD)

One of the leading causes of death worldwide, heart disease has clear, modifiable risk factors and therefore is preventable. Stroke, type-2 diabetes, and hypertension also are cardiovascular disorders, affecting the heart and arteries. Metabolic Syndrome (or cardiometabolic syndrome) is characterized by abdominal obesity, high cholesterol, high blood pressure, and elevated fasting glucose. Although each of these symptoms is a risk factor for CVD, all can be improved by the same diet!

Dr. Caldwell Esselstyn, a cardiac surgeon recognized for his work at the Cleveland Clinic in Ohio, published a study in 2014 demonstrating that diet modification alone can serve as a powerful agent for disease reversal. Of the 198 patients with established CVD, 177 adhered to a low-fat vegan diet. Upon follow-up (an average of 3.7 years later), it was established that only one person out of the 177 (0.6 percent) had a recurring CVD event. During the same period, a staggering 13 of 21 persons (62 percent) in the control group—those who did not adhere to a plant-based diet—suffered a recurrence. That's a one-hundred-fold difference in outcomes between those on a low-fat vegan diet and those who did not follow a plant-based diet!

Kaiser Permanente, the largest healthcare consortium in the US–comprising nine million members, 37 medical centres, and over 600 medical offices–recently advised its 17,000 salaried physicians to recommend plant-based diets as the standard therapy for most major, chronic diseases.

Citing increasing costs and increasing rates of disease, Kaiser Permanente, the largest healthcare consortium in the US–comprising nine million members, 37 medical centres, and over 600 medical offices–recently advised its 17,000 salaried physicians to recommend plant-based diets as the standard therapy for most major, chronic diseases. Rather than prescribing cholesterol-lowering drugs and other pharmaceuticals, doctors increasingly are providing lifestyle and dietary counselling to their patients as a first line of treatment. Private insurance companies are beginning to take notice.

THE BENEFITS OF EATING PLANTS

Nutrient Density

Most plant foods are extremely nutrient-dense, offering many vitamins and minerals for relatively few calories. Since fruits and vegetables contain fewer calories and more nutrients per serving than meat does, comprising the bulk of your diet with plant foods will significantly increase the amount of vitamins and minerals you consume. Even iceberg lettuce—often a token accompaniment to the main dish, and commonly regarded as "only water"—provides significant amounts of iron, calcium, vitamin A, vitamin K, and 7 grams of protein per 100 calories. Kale is even more nutrient-dense.

Fiber

Fiber is found in plant leaves, stems, and seeds. It is only found in plant foods and has no calories, despite being listed under "carbohydrate" on nutrition labels. Both soluble and insoluble fiber produce a sensation of fullness relatively quickly and leave one feeling full longer. A study in New Zealand analyzed dietary patterns involving three components of food choices--healthful, low-cost, and environmentally sustainable. It found that high-fiber foods (vegetables and whole grains) ranked the highest when all three categories were considered together. Fiber has a number of functions in our digestive system, from slowing gastric emptying to easing waste elimination. Generally, the greater a person's fiber intake (found only in plant foods), the less he or she is at risk for a number of diseases, including coronary heart disease, type-2 diabetes, and cancer. The US recommendation for fiber intake is 25 - 35 grams per day. Unfortunately, the average American eats only half this amount.

Phytochemicals and Antioxidants

Phytochemicals—nutrients that are present in fruits, vegetables, grains, legumes, nuts, seeds, spices, herbs, tea, and coffee—are widely thought to safeguard and to enhance human health. Among these nutrients are vitamins and minerals such as beta-carotene, vitamins C and E, copper, zinc, and selenium, all of which are also antioxidants. By scavenging the body for damage-inducing free radicals that may harm cell membranes and DNA molecules, antioxidants contribute toward the prevention or reduction of age-related

People who follow whole-food, plant-based diets have higher intakes of phytochemicals and antioxidants, a fact reflected by their greater plasma antioxidant levels.

diseases (such as macular degeneration), cellular diseases (such as cancer), and cardiovascular diseases (heart disease).

People who follow whole-food, plant-based diets have higher intakes of phytochemicals and antioxidants, a fact reflected by their greater plasma antioxidant levels. This may partially explain the lower rates of chronic diseases found in vegetarians and vegans.

In the last few decades increased knowledge of phytochemicals has led to significant research to identify and determine their benefits. It is the future of nutrition and one can only hypothesize that as we learn more about phytochemicals, we will further substantiate the benefits of plant-based diets.

The best way to obtain antioxidants is directly through whole plant foods. There are many classes of phytochemicals. Some common examples and their food sources are listed below.

Classes of Phytochemicals	Sources
Carotenoids	dark green leafy vegetables; red, yellow, and orange vegetables; cantaloupe; watermelon; and citrus fruits
Flavonoid	Cacao; berries; grapes; citrus fruits; onions; broccoli; cranberries; peanuts; cinnamon; green and black teas
Phytoestrogen	soybeans and soy products (e.g. tofu and tempeh); flaxseeds, sesame seeds; and rye
Sulfides and Thiols	onions, garlic, and scallions
Phytosterols	wheat and wheat germ; corn; soy; peanuts; black and green teas; vegetable oils; green and yellow vegetables
Isothiocyanates	broccoli, cauliflower, kale, Brussels sprouts

Protecting the Endothelium

The endothelium is composed of the thin layer of cells that line the blood vessels. Functioning as a barrier between the bloodstream and the surrounding tissue, it is involved in vasoconstriction, vasodilation, and blood clotting. The endothelium can be damaged by chronic disorders such as hypertension, diabetes, high cholesterol, and atherosclerosis (hardening of the arteries, caused by a buildup of plaque). Endothelium dysfunction leads to oxidative stress, which, in turn, worsens these effects. Some researchers believe that phytochemicals, especially flavonoids, have an anti-inflammatory effect, by reducing platelet aggregation and plaque buildup and thus repairing the endothelium.

Healthful, Naturally Occurring Fats

Fat is required for the proper absorption of fat-soluble vitamins. The polyunsaturated fatty acids omega-3 and omega-6 (also known as essential fatty acids, or EFAs) are necessary nutrients. Even small amounts of healthy fat will help us to feel satiated, and also will enhance the flavor of food. Many plant foods that ordinarily are not associated with fat–such as greens, beans, and grains–actually do contain small amounts. In fact, most plant foods contain unsaturated fats, which may play a role in disease prevention. For instance, monounsaturated fatty acids are associated with reduced LDL ("bad") cholesterol and total cholesterol levels. Olives and olive oil, avocadoes, seeds, and some nuts contain these beneficial fats.

In fact, most plant foods contain unsaturated fats, which may play a role in disease prevention.

Polyunsaturated fats, which also are found in seeds and nuts (especially walnuts), as well as in legumes and grains, also serve to reduce LDL cholesterol and may raise HDL ("good") cholesterol too. However, saturated fats—abundant in palm oil and coconut oil—usually are not recommended, because of their link to increased LDL cholesterol levels.

Volume of Food

Since plant foods provide more volume and more nutrients for fewer calories than do meat and dairy products, switching from a SAD regimen to a plant-based diet commonly results in weight loss. The illustrative recipes in Chapters 3 - 6 include Nutritional Benefits comparison charts which show that most vegan versions of meat- and dairy-based comfort foods are more abundant while containing fewer calories.

Isolation versus Synergy

Because phytochemicals, antioxidants, and other plant-food components contribute significantly to our health, it is logical to assume that isolating these elements in the form of dietary supplements would provide similar benefits. Unfortunately, such elements in isolation do not yield the same effects. This is due to the synergy that occurs when whole plant foods are metabolized in our bodies. There are cofactors which aid in manifesting the

Consequently, the best way to derive maximum nutritional advantage is to eat a wide variety of whole plant foods.

beneficial effects of phytochemicals in whole plant foods. Consequently, the best way to derive maximum nutritional advantage is to eat a wide variety of whole plant foods.

THE "INCOMPLETE PROTEIN" MYTH

In her 1971 book, *Diet for a Small Planet*, Frances Moore-Lappé stated that a vegan diet may result in insufficient protein intake unless one ensures that certain protein sources are combined in a precise manner. Despite her having corrected this misperception in a subsequent (1981) edition of her book, the original notion of "complementarity" seems to persist in the public mind. The Academy of Nutrition and Dietetics since has corroborated Lappé's amended view, concluding that it is "totally unnecessary" to complement proteins during meals. It states:

> *[A]ppropriately planned vegetarian diets, including total vegetarian or vegan diets, are healthful, nutritionally adequate, and may provide health benefits in the prevention and treatment of certain diseases. Well-planned vegetarian diets are appropriate for individuals during all stages of the life cycle, including pregnancy, lactation, infancy, childhood, and adolescence, and for athletes.*

VEGAN ATHLETES

Since athletes' caloric needs greatly exceed those of the general population due to the energy required for training and competition, vegan athletes traditionally have been regarded as anomalies. It was believed that their diets could not supply adequate nutrition to meet their desired goals. However, whole-food, plant-based diets may yield performance-enhancing benefits superior to those offered by diets based on meat and dairy products. While many of the increasing number of athletes who choose vegan diets do so for ethical or environmental reasons, others cite health and athletic performance as reasons for their decision.

However, whole-food, plant-based diets may yield performance-enhancing benefits superior to those offered by diets based on meat and dairy products.

Protein Requirements for Vegan Athletes

The American College of Sports Medicine, in collaboration with the Academy of Nutrition and Dietetics (AND), declares that "[t]he protein recommendations for endurance and strength-trained athletes range from 1.2 to 1.7 grams per kg body weight (0.5-0.8 grams per pound [of] body weight). These recommended protein intakes can generally be met through diet alone, without the use of protein or amino acid supplements."

An 80-kilogram male athlete who is not undergoing training requires about 2,000 calories daily. Based upon the AND's recommended protein requirement of 0.8 grams per kg of body weight, this would entail 64 grams of protein per day, which can easily be met without the use of supplements. If he prepares for an athletic endurance event, his protein requirement increases to 96 grams daily. In consuming more calories to facilitate his training, he also will be acquiring the necessary additional protein. However, a strength athlete may not consume as many calories as does an endurance athlete and therefore must ensure that he consumes protein-dense foods.

> *Acquiring these levels of protein with plant foods requires little effort.*
>
> *Great sources of high-protein plant foods include beans, soy, peanuts, whole grains, nuts, and seeds.*

Acquiring high levels of protein from plant foods requires little effort. For instance, lentils and soy milk have protein contents of over 30 percent. Even some "high-carb" foods contain fair amounts of protein; 15 percent of the calories in whole-wheat pasta and 8 percent of the calories in brown rice consist of protein.

Every whole food contains protein, and therefore, essential amino acids. Even bananas and other fruits contain protein, although their percentages may be low. Great sources of high-protein plant foods include beans, soy, peanuts, whole grains, nuts, and seeds. The key is to eat a variety of foods throughout the day and to get enough calories for your training needs.

Vegan athletes should pay particular attention to acquiring sufficient calories. Healthy fats, such as those found in walnuts and almonds, seeds (flax and chia), whole soy foods, and avocados, should be part of a vegan athlete's diet.

Benefits of Eating More Vegan Foods While Training

Because easily digestible, carbohydrate-rich foods with some protein content are the best sources of nutrition during workouts, most of the food offered to marathon competitors is plant-based. Furthermore, plant foods assist in the repair of damaged cells after a demanding workout. Concentrations of the phytochemicals and antioxidants that support recovery are to be found in nearly all plant foods. Thus, every athlete could benefit from plant-based recovery meals.

Eating the best foods for your training also is beneficial to your long-term health. A breakfast of 4 mashed bananas, 1 to 2 tablespoons of almond butter, and 1 diced apple is both nutrient- and calorie-dense, and will provide energy for your training workout.

EXAMPLES OF VEGAN ATHLETES

Ultramarathon runners

Scott Jurek, the author of *Eat & Run*, is a vegan and a renowned ultra-runner who won the Western States 100-mile (161-km) race an astonishing seven times in a row. Additionally, he is a two-time winner of the gruelling 135-mile (217-km) Badwater Ultramarathon, a run through California's Death Valley in the summer, when temperatures can reach 130 °F (54 °C). His book speaks of the importance of choosing foods to fuel the body to achieve set goals.

It is evident that plant foods provide all the protein and essential amino acids that humans, including elite athletes, need.

Another ultra-runner is Catra Corbett, a vegan for over 20 years, who has competed in over a hundred 100-mile (160-km) races and events.

It is evident that plant foods provide all the protein and essential amino acids that humans, including elite athletes, need.

Mixed Martial Artists

Mixed Martial Artist (MMA) Mac Danzig has long been both a successful fighter and a proponent of veganism. He makes it very clear that his diet is about ethics. He won the King of the Cage Lightweight Championship and successfully defended it four times.

Jake Shields, who holds numerous championship belts and has been ranked among the world's top 10 middleweight fighters, also is vegan.

Strength Athletes

Strongman Patrik Baboumian, a former bodybuilder, now competes in strongman events that include log lifting, deadlifting, and bench pressing. He has squatted 272 kg (600 lb), bench-pressed 200 kg (440 lb), and holds the current keg lift world record. After transitioning from vegetarian to vegan in 2011, Patrik set a world record by carrying a 550-kg (1213-lb) yoke for 10 meters (33 ft) in 2013.

Patrik set a world record by carrying a 550-kg (1213-lb) yoke for 10 meters (33 ft) in 2013.

ASPECTS OF A WHOLE-FOOD, VEGAN DIET

Nuts

There is some controversy regarding the quantity of nuts that one ideally should consume. As nuts are small but high in calories, the impulse to overeat poses a problem. Some dietary regimens—for instance, Rip Esselstyn's Engine 2 Diet, which is designed to address obesity, heart disease, diabetes, and other disorders—recommend that nuts be avoided or eaten only sparingly. The reasoning is that the same nutrients present in nuts can be obtained from lower-calorie plant foods. Yet researchers from Loma Linda University and Dr. Walter Willett from Harvard University argue that the polyunsaturated fats, phytochemicals (including polyphenols), and vitamin E found in nuts and seeds confer health benefits, hence nut consumption should be included. Walnuts are especially high in omega-3 fatty acids, which research shows are effective in reducing the risk of cardiovascular diseases. Therefore, the decision as to whether to include nuts in one's diet should be based upon a consideration of personal factors. People who have attained their targeted weight and cholesterol level may enjoy small servings of nuts for the variety and nutritional benefits that they offer. Other individuals may wish to incorporate nuts into their diets once they have reached their initial goals. Nuts and seeds add nutrition and texture to many cooked or baked dishes, and even to smoothies.

Oils in the Vegan Diet

Vegetable oils, such as olive oil, are not whole foods because, once the oil has been extracted, the remainder of the plant—including fiber and nutrients— is discarded. Oil is 100 percent fat. Consequently, some nutritionists argue that even olive oil has no place in a disease-fighting diet. Yet olive oil has long been regarded as a health food, and there is some evidence that monounsaturated fatty acids may improve cholesterol levels in some populations. Furthermore, olive oil contains vitamin E and phytochemicals such as biophenols. The question is whether olive oil provides benefits not already offered by a whole-food, plant-based diet. It probably does not. Moreover, all oils, like nuts, are extremely calorie-dense. Thus, incorporating oil in a recipe will increase the meal's total caloric content and may impede the weight-loss benefits of a plant-based diet.

If the inclusion of small amounts of olive or canola oil renders food more appealing, the resulting increase in vegetable consumption may warrant the use of these oils.

Other nutritionists maintain that a healthful, disease-preventing diet may include small amounts of olive or canola oil. Canola oil is favored by some because of its better omega-6 to omega-3 ratio as compared to that of olive oil. Chefs claim that using small quantities of a wholesome oil to sauté vegetables significantly enhances the dishes' flavors. If the inclusion of small amounts of olive or canola oil renders food more appealing, the resulting increase in vegetable consumption may warrant the use of these oils. An awareness of their caloric density should serve as your guide in determining the quantity incorporated into your diet.

An awareness of their caloric density should serve as your guide in determining the quantity incorporated into your diet.

SOY FOODS

Including Soy Products in a Healthy Diet

Soy is a legume widely grown for its edible bean, from which tofu and several other food products are made. Nonetheless, soy has generated much controversy. One of the myths being circulated about this staple —which has long been eaten worldwide—is that its consumption triggers breast cancer, and "feminizes" men. Since 1990, over 10,000 peer-reviewed studies of soy have been conducted. The evidence indicates that eating a moderate amount of soy is safe and even beneficial.

Myth: Soy Contains Estrogen

The benefit of soy—as well as the reason for its criticism—is a phytochemical group named "isoflavones." Phytochemicals have been studied extensively, to determine their role in the prevention of chronic disease. Soy isoflavones are called phytoestrogens (plant estrogens), because they bind to some estrogen receptors in the body. Yet the activity of phytoestrogens is not the same as that of estrogen; in some cases, phytoestrogens actually have anti-estrogenic effects. Moderate soy consumption does not raise the risk for hormone-related cancers, such as breast and prostate cancer. While one published case study does relate reduced testosterone to soy consumption, the study's subject ate an unusually high amount of soy—over 14 servings a day—and his testosterone levels returned to normal once his soy intake was reduced to moderate amounts. (A

Yet the activity of phytoestrogens is not the same as that of estrogen; in some cases, phytoestrogens actually have anti-estrogenic effects.

serving of soy consists of 1 cup, or 250 ml, of soymilk; or one-half cup, or 125 ml, of tofu, tempeh, or soybeans.) Studies employing up to six servings of soy per day showed no feminizing effects, nor any effect upon men's sperm quality or quantity.

Myth: Soy Contains "anti-nutrients"

Like other wholesome plant foods such as whole wheat and potatoes, soy contains phytates, which may reduce the absorption of some minerals. However, this does not lead to any vitamin or mineral deficiencies in a healthful, balanced diet. For instance, the calcium in soy milk is absorbed at about the same rate as that of cow's milk. Ferritin, the type of iron present in high amounts in soy, is absorbed especially well by individuals with low iron stores. Firm tofu is an excellent source of iron. The addition of food rich in vitamin C, such as citrus fruits or bell peppers, significantly increases iron absorption.

Myth: Only Fermented Soy Foods are Healthful

Much of the evidence regarding soy's health benefits, including lower risks of breast cancer and of heart disease, comes from Asian populations, which have been consuming soy habitually for centuries. Some critics argue that soy is merely a condiment in most Asian cultures, or that only fermented soy is eaten, but these contentions are not supported by facts. While fermented soy foods, such as natto and miso, are consumed regularly in Japan, the same is true of tofu. Moreover, most of the soy eaten in China is not fermented. In Indonesia, where tempeh (a fermented soy and grain cake) is popular, non-fermented types of soy also are eaten regularly. The total amount of phytochemicals in fermented versus non-fermented soy (such as tofu, edamame, and soy milk) varies little; there is no objective reason to prefer one type over another.

Myth: Soy Causes Cancer

Despite considerable confusion as to whether the isoflavones found in soy play any role in the incidence of cancer—particularly breast cancer—the latest research indicates that soy's effect is either neutral or protective. Previously, many believed that soy should be avoided by those recovering from estrogen-receptor-positive breast cancer. To the contrary, the most current research shows that isoflavones have anti-estrogen activity in reproductive cells, resulting in a beneficial outcome.

...the most current research shows that isoflavones have anti-estrogen activity in reproductive cells, resulting in a beneficial outcome.

Soy Products

Traditional, non-GMO soy foods such as tofu, tempeh, edamame, and soy milk are preferable to more processed versions like the soy protein isolate so prevalent in many faux meats. However, eating faux meats occasionally does not pose a health risk, despite some products containing higher sodium content or added fat. Individuals watching their sodium or fat intake may wish to note the nutritional labelling on these products. At the same time, one should not become reliant upon soy, but should eat a variety of legumes, which also includes beans, lentils, split peas, and chickpeas.

Soy Benefits

Clinical studies suggest that soy may reduce menopausal symptoms and may lower the risk of heart disease and osteoporosis. The US Food and Drug Administration allows soy products to carry a "Heart Healthy" label. This signifies that one serving (slightly over 6 grams of soy protein) may reduce cholesterol levels when consumed daily. The American Heart Association recommends 25 grams of soy protein daily, as part of a diet low in saturated fat and cholesterol. This amount of soy protein is equivalent to 8 oz (225 grams) of firm tofu, or 2.5 cups (625 ml) of soy milk.

CONCLUSION

Although nutritional research is a continual process, decades of studies demonstrate that plant-based diets including whole grains, vegetables, legumes, and fruits contribute greatly toward the prevention of many chronic diseases. What's more, the diet that is most beneficial to your short- and long-term health also will generate the best possible outcome for our planet and all its inhabitants.

Chapter 3
How Veganizing Beef Recipes Improves Environmental and Nutritional Parameters: *An Analysis*

These next few chapters illustrate the impacts that our food choices at every meal have upon the environment and our health.

I created vegan versions of traditional recipes for some popular North American comfort foods. Then the *differences* for 3 environmental and 7 nutritional parameters, between the meat and vegan versions of the recipes are calculated. The findings are presented in the Environmental Savings Tables and Nutritional Benefits Charts.

The Environmental Savings Table shows typical savings achieved with respect to water footprint, amount of manure produced, and greenhouse gas emissions (in terms of CO_{2e}). The Nutritional Benefits Chart gives the benefits in total fat, saturated fat, cholesterol, iron, sodium, protein, and total calories.

Notes:

There are many "meat-like" vegan products on the market. Various companies use seitan (wheat gluten) or soy, or a combination of the two. As there is insufficient information to analyse the environmental impacts for specific brands of vegan "meats", organic tofu is used for the analyses. This represents a close approximation to the main ingredient in popular meat substitutes.

Meat alternatives now range from beef-like to chicken-like counterparts and usually can be directly substituted, either as an ingredient in a traditional recipe or as an entrée in its own right. I recommend non-GMO soy and organic tofu, where possible. Depending on the coagulant used—calcium chloride, calcium sulfate, or magnesium sulfate—tofu can be beneficially high in calcium or magnesium.

Firm or extra-firm tofu is available in packages of different sizes, ranging from 350 grams (12.5 oz) to 425 grams (15 oz). To render as exact a comparison of environmental and nutritional benefits as possible, an equivalent amount of tofu is substituted for the meat in each recipe.

What's the Beef?

ENVIRONMENT

As noted in Chapter 1, Figure 2, beef has the highest average water footprint of all the animal products commonly consumed: 15,415 liters per kg of beef. This figure is almost twice that of lamb or goat at 8,763 liters. Although nuts have the second highest water footprint, people generally do not eat nuts in the same amount or to the same degree as they do animal flesh. Additionally, the unit costs for nuts are much higher than that for animal flesh.

Of all the farmed animals, cattle produce the highest amount of excrement per kg of meat yielded: 39 kg per kg of meat (Chapter 1, Figure 5). The livestock industry maintains that animal manure is composted and is applied as fertilizer to agricultural land. Nonetheless, manure is a significant source of viruses, bacteria, heavy metals, hormones, antibiotic residues, and pesticides, not to mention prions, the agents of Mad Cow Disease. While composting may destroy some bacteria and viruses, it does not remove or denature such contaminants as heavy metals, drug residues, pesticides, and prions. Since health concerns make it illegal to dispose of human waste by spreading it on agricultural land, it is astounding that the disposal of animal waste in the current manner is common industry practice.

There is no practical way to dispose of pathogenic, contaminated manure safely without generating public health problems. Manure is a water pollutant; its application usually exceeds the soil's ability to accommodate it, thus resulting in the pollution of ground and surface waters. The land cannot absorb the nutrients and denature the bacteria rapidly enough.

Conversely, vegan-organic, or "veganic" agriculture (which uses plant-only compost materials, also referred to as "stock-free," or "plant-based") has been practiced for centuries. While evidence for such large-scale agriculture is lacking, it is self-evident that all nutrients necessary for plant growth originate from the soil or are manufactured by plants. Just as humans do not need to obtain protein by cycling plant proteins through animals, crops do not need to obtain their nutrients via the digestive process of animals in the form of manure.

Just as humans do not need to obtain protein by cycling plant proteins through animals, crops do not need to obtain their nutrients via the digestive process of animals in the form of manure.

Interestingly, lamb has the highest source of GHGs, at 39.2 kg of CO_{2e} per kg of meat (Chapter 1, Figure 1). Cows, the second highest source, produce 27 kg of CO_{2e} per kg of beef. Environmentally speaking, the true cost of beef is far greater than its unit price at the grocery store. What if beef producers were required to incorporate the exorbitant environmental costs of water usage and greenhouse gas emissions of their products, and to pass these on to consumers? Might this be fairer to people who do not eat beef but who nonetheless bear the consequences of environmental degradation?

What if beef producers were required to incorporate the exorbitant environmental costs of water usage and greenhouse gas emissions of their products, and to pass these on to consumers?

In a landmark study in the Proceedings of the National Academy of Sciences, a uniform methodology was used to compare the environmental footprint of major animal-based industries in the US. Professor Gidon Eshel and his team found that beef production requires 28 times more land, 11 times more irrigation water, produces 5 times more greenhouse gases, and 6 times more reactive nitrogen fertilizer than do other US livestock groups.

HEALTH

A 2013 study by ML McCullough, et. al., published in the *Journal of Clinical Oncology*, examined the effect between red and processed meat consumption on colorectal cancer survivors. It found that those survivors who subsequently ate the most red or processed meats were more likely to die over a 7.5-year period than those who ate the least. Moreover, they found a 29 percent higher risk of death from all causes, and a 63 percent higher risk of death from heart disease in the group that ate the most red and processed meat as compared to those who ate the least.

Colorectal cancer survivors who ate the most red or processed meats had a 29 percent higher risk of death from all causes, and a 63 percent higher risk of death from heart disease than those who ate the least.

Women who consumed the most red meat were 22 percent more likely to develop breast cancer, as compared to those who ate the least.

Similarly, a 2014 study by MS Farvid, published in the *British Medical Journal,* found that women who consumed the most red meat were 22 percent more likely to develop breast cancer, as compared to those who ate the least. Each additional serving per day resulted in a 13 percent increased risk for breast cancer. Participants were premenopausal women in the Harvard Nurses' Health Study II.

Another 2013 study by A Pan, et. al., published in the *Journal of the American Medical Association,* indicates that increases in the consumption of red meat contribute to weight gain and to increased risk for diabetes. An increase of only slightly more than half a serving of red meat per day increases the risk for type 2 diabetes by 48 percent!

An increase of only slightly more than half a serving of red meat per day increases the risk for type 2 diabetes by 48 percent.

Every 50 gram (less than 2 oz) serving, such as a regular-sized hot dog, increased heart failure risk by 8 percent and the chances of dying from heart failure by 38 percent.

Furthermore, a 2014 study of Swedish men by J. Kaluza, et. al., published in *Circulation: Heart Failure*, found that processed meats may double your risk of dying from heart failure over a 12-year period. Every 50 gram (less than 2 oz) serving, such as a regular-sized hot dog, increased heart failure risk by 8 percent and the chances of dying from heart failure by 38 percent. These are alarming statistics.

"I am in favor of animal rights as well as human rights.
That is the way of a whole human being."

~ Abraham Lincoln

Tex-Mex Tortillas with a Kick (6 servings)

Beef Recipe	Vegan Version
6 tortillas	6 tortillas
1 Tbsp oil	1 Tbsp (15 ml) oil
1 onion, chopped	1 onion, chopped
4 garlic cloves, minced	4 garlic cloves, minced
1 lb (450 g) ground beef	*3 cups (750 ml) of pinto beans*
½ tsp black pepper	½ tsp (3 ml) black pepper
2 tsp chili powder	2 tsp (10 ml) chili powder
1 cup corn	1 cup (250 ml) corn
1 cup pasta sauce	1 cup (250 ml) pasta sauce
1 cup pineapple salsa	1 cup (250 ml) pineapple salsa
1 tsp Tabasco sauce	1 tsp (5 ml) Tabasco sauce

Instructions:

1. In a skillet, heat the oil and add the onion and garlic. Cook on medium heat for 5 minutes, until the onion is softened.
2. Add the beans, spices, corn, pasta sauce, salsa, and Tabasco. Bring to a boil then reduce to simmer for 5 to 10 minutes. Let sauce thicken.
3. Spoon the mixture into tortillas and wrap.

FACT: Did you know that the caloric content of the food (mostly grains) fed to the global cattle population of 1.3 billion can feed 8.7 billion people?

The current global population is 7 billion.

Imagine the amount of food that can be made available to feed the hungry if people were to stop eating cows.

Environmental Savings

	Beef Recipe 1lb beef (450 grams)	**Vegan Version** 3 cups cooked* pinto beans (750ml)	**Total Savings** (Beef minus Vegan)	**Savings Per Serving** (Beef minus Vegan)
H_2O	6,937 liters (1,833 US gal)	912 liters (241 US gal)	6,025 liters (1,592 US gal)	**1,004 liters (265 US gal)**
M	17.5 kg (38.5 lbs)	0	17.5 kg (38.5 lbs)	**2.9 kg (6.4 lbs)**
GHG	12.2 kg CO_{2e}	.45 kg CO_{2e}	11.7 kg CO_{2e}	**2 kg CO_{2e} <=> (Driving 7.7 km or 4.8 miles)**

**1 cup dry beans weighs 8 oz (225 grams), yields 3 cups cooked*

H_20 = Water Footprint, M =Manure produced, GHG = Greenhouse Gas

Nutritional Benefits per Serving

	Beef Recipe	Vegan Version
Calories	370	260
Protein (g)	17	10
Fat (g)	21	7
Saturated fat (g)	7	1
Cholesterol (mg)	55	0
Iron (mg)	2.2	2.5
Sodium (mg)	640	590

**It is recognized that cattle raised in different parts of North America require different amount of water. The water footprint data represent global averages, and the reference for comparison remains consistent; it is the relative difference between the animal and the plant products that is being examined.*

Quick Chili (4 servings)

Beef Recipe	Vegan Version
1 lb (450 grams) ground beef	*2 cups (500 ml) cooked kidney beans*
½ cup kidney beans	Additional ½ cup (125 ml) cooked kidney beans
1 medium onion, chopped	1 medium onion, chopped
4 garlic cloves, minced	4 garlic cloves, minced
1 small green pepper, chopped	1 small green pepper, chopped
2 cups canned diced tomatoes	2 cups (500 ml) canned diced tomatoes
½ tsp salt	½ tsp (3 ml) salt
½ tsp pepper	½ tsp (3 ml) pepper
½ tsp cayenne	½ tsp (3 ml) cayenne
1 Tbsp chili powder	1 Tbsp (15 ml) chili powder
1 tsp oregano	1 tsp (5 ml) oregano
2 Tbsp oil	2 Tbsp (30 ml) oil

Instructions:

1. In a large pot, sauté the onion and garlic in oil for a few minutes, until onion is soft.
2. Add the remaining ingredients, and bring to gentle boil. Then decrease the heat and allow to simmer for 15 minutes

FACT: Cows are intelligent animals that form social hierarchies and lifelong bonds. They can recognize up to 100 individuals, and also will let a farmer know when he's late for their feeding.

A cow's natural life span is approximately 20 years; yet beef cattle are killed at the age of only 18 months.

Environmental Savings

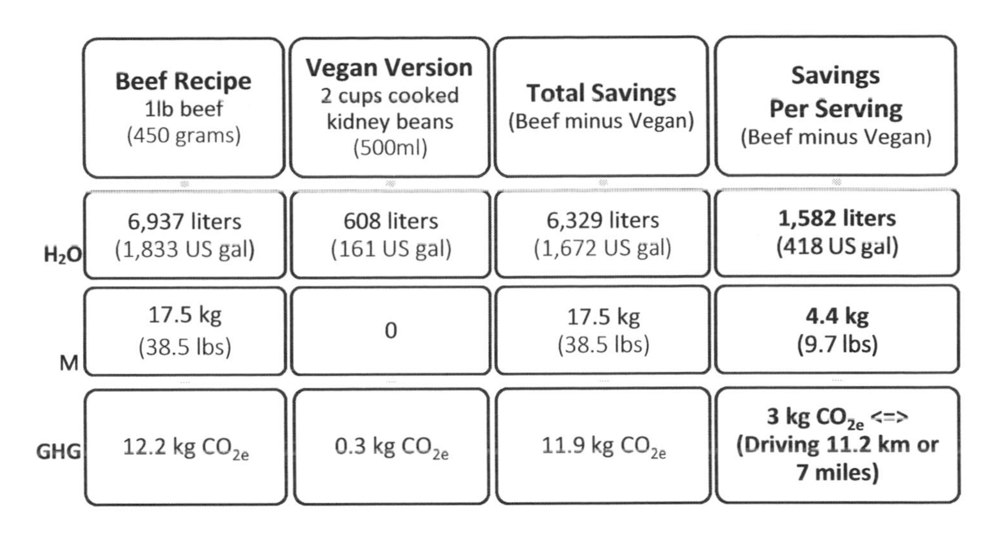

	Beef Recipe 1lb beef (450 grams)	**Vegan Version** 2 cups cooked kidney beans (500ml)	**Total Savings** (Beef minus Vegan)	**Savings Per Serving** (Beef minus Vegan)
H_2O	6,937 liters (1,833 US gal)	608 liters (161 US gal)	6,329 liters (1,672 US gal)	**1,582 liters (418 US gal)**
M	17.5 kg (38.5 lbs)	0	17.5 kg (38.5 lbs)	**4.4 kg (9.7 lbs)**
GHG	12.2 kg CO_{2e}	0.3 kg CO_{2e}	11.9 kg CO_{2e}	**3 kg CO_{2e} <=> (Driving 11.2 km or 7 miles)**

H_2O = Water Footprint, M =Manure produced, GHG = Greenhouse Gas

Nutritional Benefits per Serving

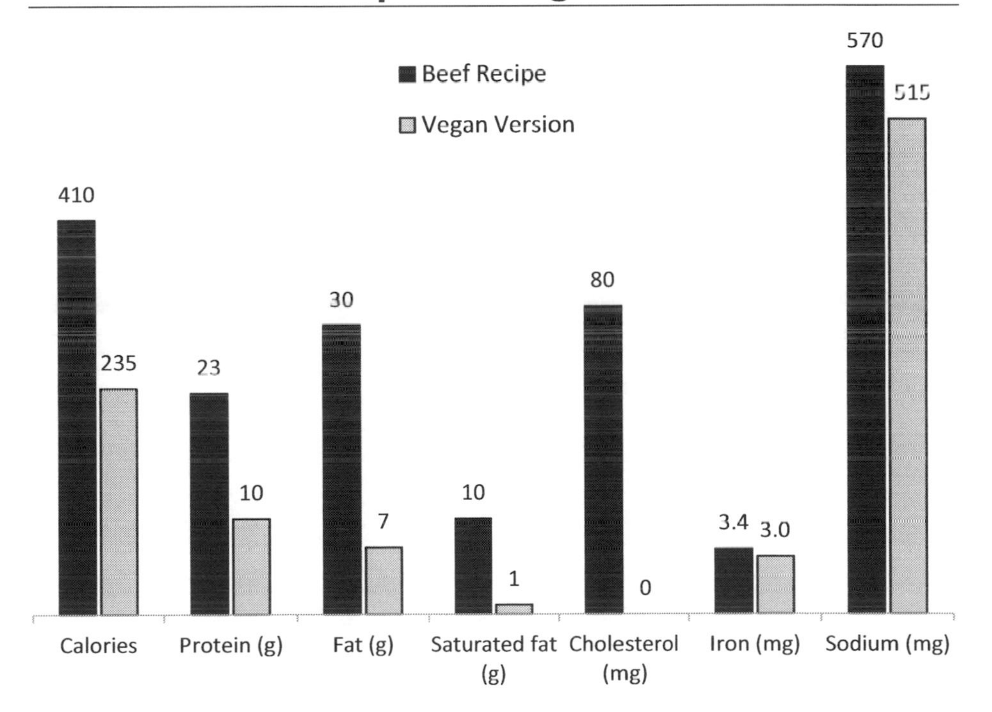

One-Pot Spaghetti (6 servings)

Beef Recipe	Vegan Version
1 lb (450 grams) ground beef	*1 lb (450 grams) extra-firm organic tofu*
1 Tbsp oil	1 Tbsp (15 ml) oil
1 onion, chopped	1 onion, chopped
4 garlic cloves, minced	4 garlic cloves, minced
2 cups pasta sauce – Marinara, basil, or mushroom	2 cups (500 ml) pasta sauce – Marinara, basil, or mushroom
½ package (6 – 7 oz) whole wheat spaghetti	½ package (6 – 7 oz, or 175 – 200 g) whole wheat spaghetti
½ tsp salt	½ tsp (3 ml) salt
1 tsp pepper	1 tsp (5 ml) pepper
1 tsp oregano	1 tsp (5 ml) oregano

Instructions:

1. Cook the spaghetti according to package directions. Drain and set aside.
2. Crumble the tofu in a blender, or use a masher to obtain pea-sized crumbles.
3. In a large pot, heat the oil and cook the onion, garlic, and tofu until onion is soft, about 5 to 7 minutes. Add salt, pepper, and oregano.
4. Stir in pasta sauce and the cooked spaghetti, and mix all ingredients well. Add more sauce if desired, or a little water while stirring, if the mixture is too thick. Heat through for another 5 minutes.

FACT: Imagine the reduction in the incidence of coronary heart disease if people stopped eating cows.

Healthcare costs associated with the treatment of these diseases would plummet.

Environmental Savings

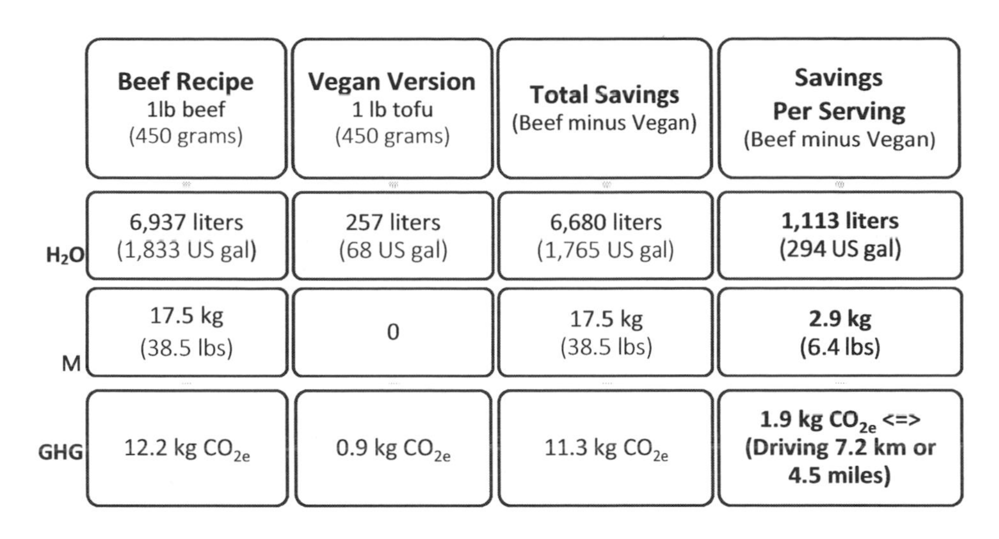

	Beef Recipe 1lb beef (450 grams)	Vegan Version 1 lb tofu (450 grams)	Total Savings (Beef minus Vegan)	Savings Per Serving (Beef minus Vegan)
H_2O	6,937 liters (1,833 US gal)	257 liters (68 US gal)	6,680 liters (1,765 US gal)	**1,113 liters (294 US gal)**
M	17.5 kg (38.5 lbs)	0	17.5 kg (38.5 lbs)	**2.9 kg (6.4 lbs)**
GHG	12.2 kg CO_{2e}	0.9 kg CO_{2e}	11.3 kg CO_{2e}	**1.9 kg CO_{2e} <=> (Driving 7.2 km or 4.5 miles)**

H_2O = Water Footprint, M =Manure produced, GHG = Greenhouse Gas

Nutritional Benefits per Serving

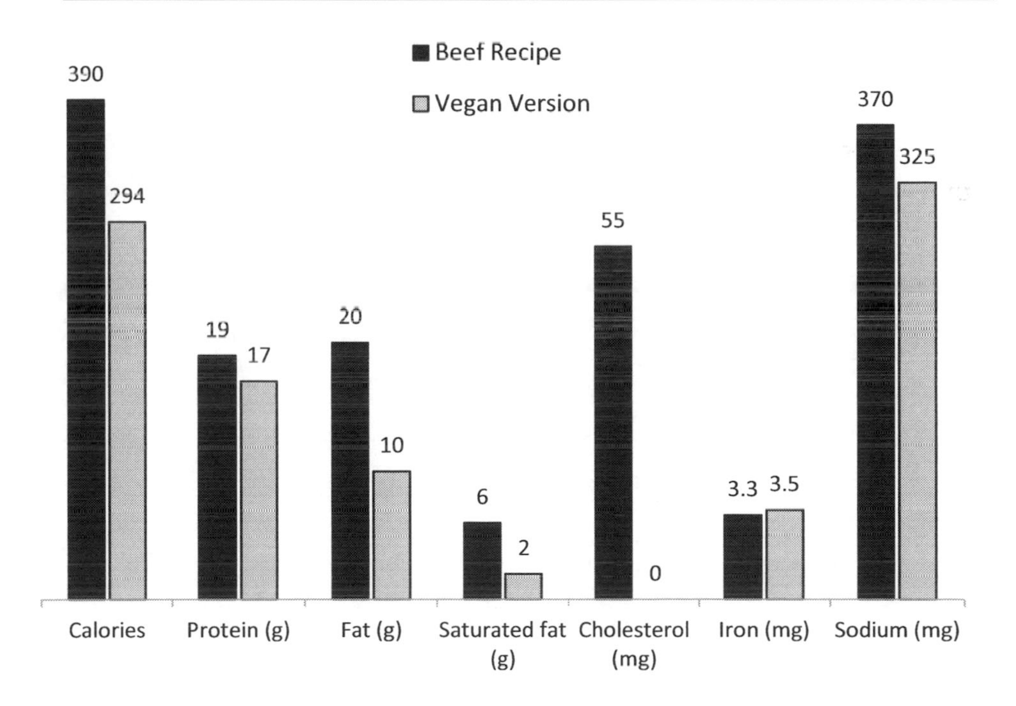

Sloppy Joes (4 servings)

Beef Recipe	Vegan Version
1 lb (450 grams) ground beef	*1 lb (450 grams) firm organic tofu*
1 Tbsp oil	1 Tbsp (15 ml) oil
1 cup pasta sauce	1 cup (250 ml) pasta sauce
2 Roma tomatoes, diced	2 Roma tomatoes, diced
1 onion, diced	1 onion, diced
2 garlic cloves, minced	2 garlic cloves, minced
1 tsp green chilies or Jalapeno pepper, diced	1 tsp (5 ml) green chilies or Jalapeno pepper, diced
4 Tbsp ketchup	4 Tbsp (60 ml) ketchup
4 Tbsp BBQ sauce	4 Tbsp (60 ml) BBQ sauce
4 whole wheat hamburger buns	4 whole wheat or gluten-free hamburger buns

Instructions:

1. Crumble the tofu in a blender, or use a masher to obtain pea-sized crumbles.
2. Heat the oil in a skillet and cook the onion, garlic, and tofu until onion is soft, about 5 minutes. Drain off excess liquid.
3. Add pasta sauce, tomatoes, chilies, ketchup, and BBQ sauce. Reduce heat to simmer for about 10 minutes, so that mixture is heated thoroughly.
4. Scoop mixture onto toasted buns.

FACT: Over 80 percent of grain-fed cattle in Canada are to be found on feedlots containing over 1,000 animals. The most modern operations have a capacity of over 40,000 cattle.

In the US, a single feeding operation in Arizona with just one feedlot has a capacity of 175,000 cattle. Some feeding operations have multiple feedlots; for example, one in Colorado has 10 feedlots, with a total capacity of 811,000 cattle.

Environmental Savings

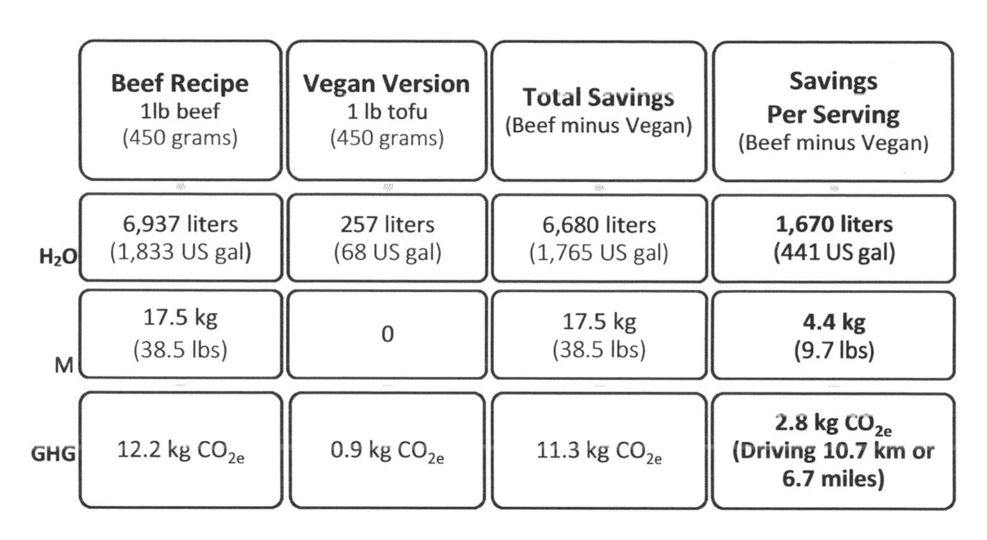

	Beef Recipe 1lb beef (450 grams)	**Vegan Version** 1 lb tofu (450 grams)	**Total Savings** (Beef minus Vegan)	**Savings Per Serving** (Beef minus Vegan)
H_2O	6,937 liters (1,833 US gal)	257 liters (68 US gal)	6,680 liters (1,765 US gal)	**1,670 liters (441 US gal)**
M	17.5 kg (38.5 lbs)	0	17.5 kg (38.5 lbs)	**4.4 kg (9.7 lbs)**
GHG	12.2 kg CO_{2e}	0.9 kg CO_{2e}	11.3 kg CO_{2e}	**2.8 kg CO_{2e} (Driving 10.7 km or 6.7 miles)**

H_2O = Water Footprint, M =Manure produced, GHG = Greenhouse Gas

Nutritional Benefits per Serving

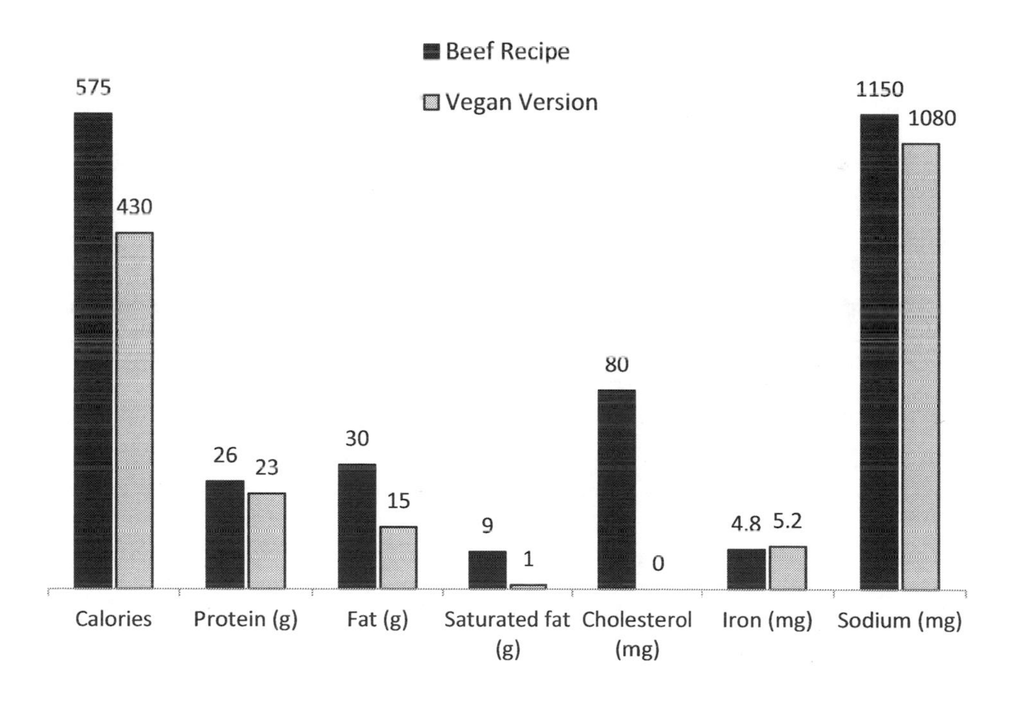

Stuffed Peppers (4 servings)

Beef Recipe	Vegan Version
2 cups cooked rice or quinoa	2 cups (500 ml) cooked rice or quinoa
1 lb (450 grams) ground beef	*1 lb (450 grams) extra-firm organic tofu*
1 Tbsp oil	1 Tbsp (15 ml) oil
1 small onion, diced	1 small onion, diced
¼ cup chopped parsley	¼ cup (60 ml) chopped parsley
¼ cup chopped basil	¼ cup (60 ml) chopped basil
½ tsp salt	½ tsp (3 ml) salt
½ tsp pepper	½ tsp (3 ml) pepper
1 tsp paprika	1 tsp (5 ml) paprika
4 large red, orange, or yellow peppers	4 large red, orange, or yellow peppers
2 cups pasta sauce, basil or mushroom flavor	2 cups pasta sauce, basil or mushroom flavor
4 oz (1 cup) shredded mozzarella cheese	*4 oz (1 cup or 250 ml) Daiya mozzarella shreds*

Instructions:

1. Cook the rice or quinoa according to package directions. Set aside.
2. Cut off bell pepper tops and discard. Clean out seeds and membranes.
3. Crumble the extra-firm tofu in food processor or use a masher to obtain pea-sized crumbles.
4. Preheat oven to 400 °F (205 °C). In a skillet, heat the oil and cook the onion until soft, about 5 minutes. Add the tofu, parsley, basil, and spices. Add half the pasta sauce and cook another 10 minutes. Remove from heat.
5. Add rice or quinoa and Daiya shreds to the skillet mixture, and combine.
6. Spoon the filling into each pepper cavity. Place peppers onto a greased baking dish. Add remaining pasta sauce to the baking dish. Cover, and bake in oven for 30 minutes. Uncover and bake for an additional 10 minutes.

Environmental Savings

	Beef Recipe 1lb beef (450 grams)	**Beef Recipe** 4 oz mozzarella cheese (0.112kg)	**Vegan Version** 1 lb tofu (450 grams)	**Vegan Version** 4 oz Daiya shreds (0.112 kg)	**Total Savings** (Beef minus Vegan)	**Savings Per Serving** (Beef minus Vegan)
H_2O	6,937 liters (1,833 US gal)	560 liters (148 US gal)	257 liters (68 US gal)	186 liters (49 US gal)	7,054 liters (1,864 US gal)	**1,764 liters (466 US gal)**
M	17.5 kg (38.5 lbs)	4.1 kg (9.1 lbs)	0	0	21.6 kg (47.5 lbs)	**5.4 kg (11.9 lbs)**
GHG	12.2 kg CO_{2e}	1.5 kg CO_{2e}	0.9 kg CO_{2e}	.65 kg CO_{2e}	12.2 kg CO_{2e}	**3 kg CO_{2e} (Driving 11.2 km or 7 miles)**

H_2O = Water Footprint, M =Manure produced, GHG = Greenhouse Gas

Nutritional Benefits per Serving

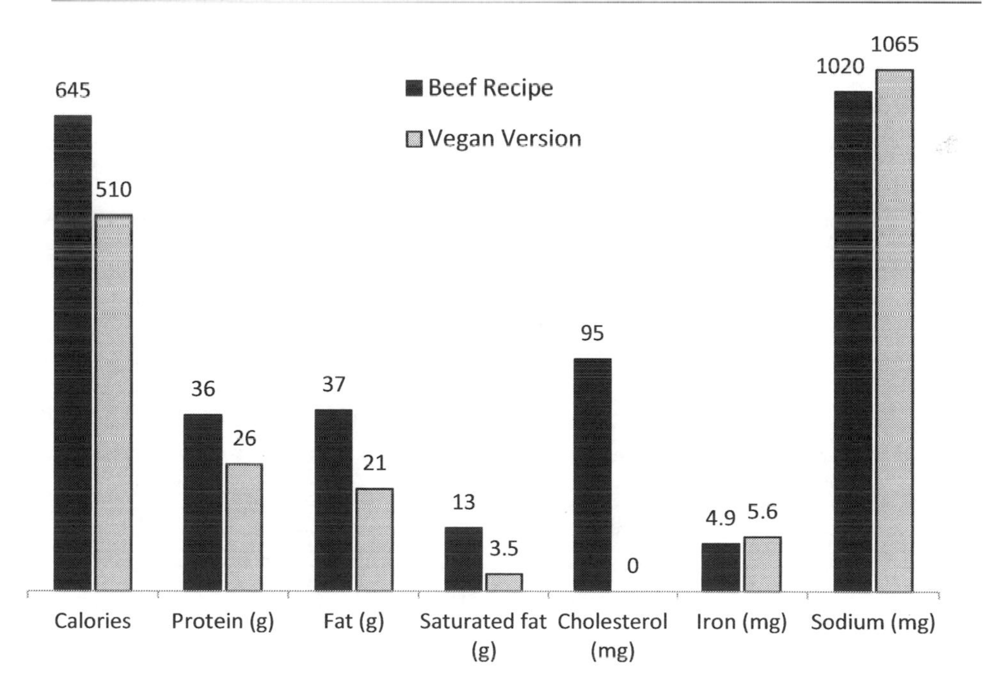

SUMMARY

Environmental Savings
(Beef minus Vegan)

Environmental Parameter per serving	Tex-Mex Tortilla	Quick Chili	One-Pot Spaghetti	Sloppy Joes	Stuffed Peppers
Water saved	1,004 liters (265 US gal)	1,582 liters (418 US gal)	1,113 liters (294 US gal)	1,670 liters (441 US gal)	1,764 liters (466 US gal)
Manure saved	2.9 kg (6.4 lbs)	4.4 kg (9.7 lbs)	2.9 kg (6.4 lbs)	4.4 kg (9.7 lbs)	5.4 kg (11.9 lbs)
GHG saved, in kg CO_{2e}	2 ⇔driving 7.7 km (4.8 miles)	3 ⇔ driving 11.2 km (7 miles)	1.9 ⇔driving 7.2 km (4.5 miles)	2.8 ⇔driving 10.7 km (6.7 miles)	3 ⇔driving 11.2 km (7 miles)

CONCLUSIONS

Environmental Savings

i. By replacing 1 lb of beef with the equivalent amount of either beans or tofu, water savings are achieved that range from 1,004 liters (265 US gallons) to 1,670 liters (441 gallons) per serving. These figures represent just one typical serving at one meal for one person. Imagine the amount of water that would be saved if cattle raising were to be eliminated!

ii. Manure generated would decrease by 2.9 to 5.4 kg per serving. That's equivalent to 6.4 lbs less manure per serving of tortilla to 11.9 lbs less manure per serving of Stuffed Peppers. This higher amount reflects the manure produced as a result of the inclusion of cheese.

iii. Greenhouse gas emissions range from 1.9 to 3 kg of CO_{2e} Using EPA's Greenhouse Gas Equivalencies Calculator, this amounts to driving the average American car between 7.2 km (4.5 miles) and 11.2 km (7 miles). All of this egregious use of resources is just for one single meal.

iv. Consuming beef and cheese together compounds the environmental impact. Since cheese has a significantly high water footprint (5,000 liters per kg) and the third highest GHG emission (13.5 kg of CO_{2e}), recipes that combine beef and cheese (as is popular in North American cuisine) are particularly detrimental to the environment. Thankfully, there are several tasty cheese alternatives on the market that are both soy- and gluten-free. While they are not necessarily lower in calories than are dairy cheeses, these alternatives are, at least, cholesterol-free and much less damaging to the environment.

SUMMARY

Nutritional Benefits
(Beef minus Vegan)

Nutritional Parameter per serving	Tex-Mex Tortilla	Quick Chili	One-Pot Spaghetti	Sloppy Joes	Stuffed Peppers
Calories	110 fewer	175 fewer	96 fewer	145 fewer	135 fewer
Protein	7 g less	13 g less	2 g less	3 g less	10 g less
Fat	14 g less	23 g less	10 g less	15 g less	16 g less
Saturated fat	6 g less	9 g less	4 g less	8 g less	10.5 g less
Cholesterol	55 mg eliminated	80 mg eliminated	55 mg eliminated	80 mg eliminated	95 mg eliminated
Iron	0.3 mg more	0.4 mg less	0.2 mg more	0.4 mg more	0.7 mg more
Sodium	50 mg less	55 mg less	45 mg less	70 mg less	45 mg more

CONCLUSIONS

Nutritional Benefits

i. Replacing beef with beans or tofu significantly reduces both calories and cholesterol, as well as the amounts of fat and saturated fat ingested. For example, the vegan chili contains 23 g less fat per serving, thus eliminating more than 1.5 Tbsp of grease. This change alone, if implemented regularly, is likely to reduce the risks of most major chronic diseases, including obesity, heart disease, stroke, diabetes, and some cancers. Beans are recognized for providing potent protection against a number of diseases. Yet, despite recommendations to increase their intake, most Americans eat less than one cup of beans weekly. Tofu often is regarded with even greater skepticism. Some of the recipes included in this book may serve to change such an opinion.

ii. Although some of the vegan versions of these recipes contain less protein than do their traditional counterparts, the meal as a whole usually provides adequate protein. For example, the chili's protein content is only 10 g, but the addition of one cup of brown rice and two cups of vegetables would result in a complete meal that supplies 23 g of protein, which is over one-third of the average daily protein requirement for most people. Other vegan recipes may furnish nearly the same protein content as do their meat and dairy counterparts.

iii. Although beef is known for its high iron content, while plant foods are perceived to be iron-poor, it is worth noting that the vegan versions contain comparable amounts as the beef recipes.

Chapter 4

How Veganizing Pork Recipes Improves Environmental and Nutritional Parameters: *An Analysis*

Pork: The *Other* Red Meat

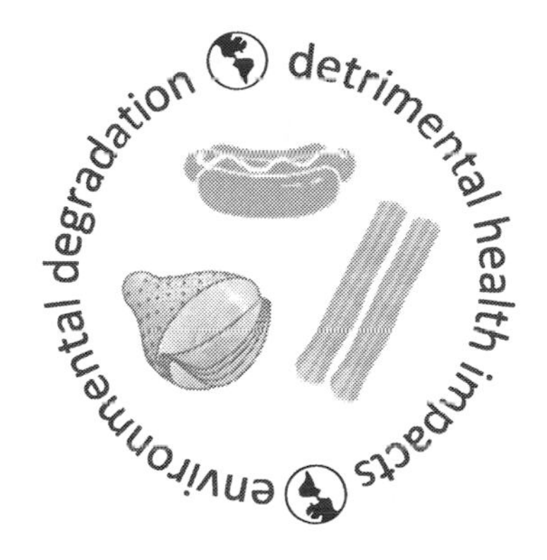

ENVIRONMENT

At about 6,000 liters per kg, the water footprint for pork (Chapter 1, Figure 2) is substantially less than that for beef (roughly 15,000 liters per kg), but still much larger than that for other animal products and certainly far greater than that for plant protein foods. It is hardly surprising that the pig produces 8.6 kg of manure per kg of meat (Chapter 1, Figure 5). Additionally, pigs' GHG emissions of 12.1 kg of CO_{2e} approach those for cheese at 13.5 kg of CO_{2e} (Chapter 1, Figure 1).

Industrial hog farms use open-air waste lagoons, often the size of several football fields, to hold the manure. Unfortunately, these lagoons are prone to leaks and spills. Chapter 1 mentioned the Carolina hog-farm lagoon incident in 1995, in which 25 million gallons of manure spilled into the New River. This resulted in the deaths of over 10 million fish, and closed over 350,000 acres of coastal wetlands to shell-fishing. Unfortunately, this was not an isolated incident. In 2011, an Illinois hog farm spilled 200,000 gallons of manure into a creek, killing 110,000 fish. Huge hog farms also emit hydrogen sulfide, a gas that can cause flu-like symptoms in people. At sufficiently high concentrations, brain damage can occur. In 1998, 19 people died as a result of hydrogen sulfide emissions from manure pits.

HEALTH

Despite the pork industry's insistent tagline of "the other white meat"—presumably intended to imply that pork is healthful—it is still red meat; therefore, its consumption should be kept to a minimum. Like all animal foods, pork is high in fat and in saturated fat, and has no fiber or significant phytochemical content. Thus, it possesses none of the traits associated with reducing the risk of chronic diseases.

Dutch scientist A.Y. Hoestra's article "The hidden water resource use behind meat and dairy" reveals that pork has the highest amount of overall fat per kg of meat (259 grams), and the lowest protein content per kg of meat produced (105 grams) among the four most commonly farmed animals: cattle, goat/lamb, pigs, and chickens. The reason we don't see that fat is that much of it has been trimmed away by the butcher before it reaches the supermarkets shelves. This means that pork contains the highest caloric count per kg of meat.

A 2012 Harvard study with lead researcher Dr. Frank Hu, Professor of Medicine, found that each daily increase of one serving (3 oz or 100 grams) of red meat was associated with a 12 percent greater risk of death. The inclusion of processed meats such as bacon in such a diet increased this risk to 20 percent. The conclusion is inescapable: the more red meat you eat, the greater the risk of dying of cancer and of cardiovascular and other diseases. These study results controlled for factors such as physical activity, smoking, higher BMI, and other variables.

The conclusion is inescapable: the more red meat you eat, the greater the risk of dying of cancer and of cardiovascular and other diseases.

Traditional southern and African-American diets tend to be high in pork. This may partially explain why these populations have especially high rates of obesity, hypertension, and heart disease.

"At the moment our human world is based on the suffering and destruction of millions of non-humans. . . [O]nce you have admitted the terror and pain of other species you will be . . . aware of the endless permutations of suffering that support our society."

- Sir Arthur Conan Doyle

Szechwan Ground Tofu with Green Beans (4 servings)

Pork Recipe	Vegan Version
1 lb (450 grams) ground pork	*1 lb (450 grams) extra-firm organic tofu, crumbled*
1 Tbsp extra light olive oil	1 Tbsp (15 ml) extra light olive oil
4 cups green beans, ends removed, and halved	4 cups (1 liter) green beans, ends removed, and halved
2 cloves garlic, minced	2 cloves garlic, minced
¼ cup hoisin sauce	¼ cup (60 ml) hoisin sauce
1 Tbsp low-sodium soy sauce	1 Tbsp (15 ml) low-sodium soy sauce
1 tsp chili pepper flakes	1 tsp chili pepper flakes

Instructions:

1. Quickly blanch or steam the green beans, allowing them to retain their crunchiness. Set aside.
2. Heat oil in a wok at medium-high heat, and stir-fry the crumbled tofu with garlic for several minutes.
3. Combine the hoisin sauce, soy sauce, and chili pepper flakes in a small bowl. Stir to mix well. Add the green beans to the wok. Pour this mixture over the tofu and green beans, stirring occasionally.
4. Cook for a few minutes, until the green beans are tender but not overcooked. Enjoy with a steamed wild-rice blend.

FACT: Pigs' cognitive abilities are greater than those of a three-year-old child, while those of dogs are only equivalent to 22-month-old infants. In 1998, Candace Croney, an Associate Professor of Animal Sciences at Purdue University, found that pigs were able to perform tasks that only Rhesus monkeys and chimpanzees previously were considered capable of accomplishing. Pigs are quick learners; they can perform tasks that respond to visual cues, and can discriminate odors such as spearmint from other mints. These social creatures can also deduce how to open gates.

Environmental Savings

	Pork Recipe 1lb (450 grams)	**Vegan Version** 1 lb tofu (450 grams)	**Total Savings** (Pork minus Vegan)	**Savings Per Serving** (Pork minus Vegan)
H_2O	2,695 liters (712 US gal)	257 liters (68 US gal)	2,438 liters (644 US gal)	**610 liters (161 US gal)**
M	3.9 kg (8.6 lbs)	0	3.9 kg (8.6 lbs)	**1 kg (2.2 lbs)**
GHG	5.45 kg CO_{2e}	.9 kg CO_{2e}	4.55 kg CO_{2e}	**1.1 kg CO_{2e} (Driving 4.2 km or 2.6 miles)**

H_2O = Water Footprint, M =Manure produced, GHG = Greenhouse Gas

Nutritional Benefits per Serving

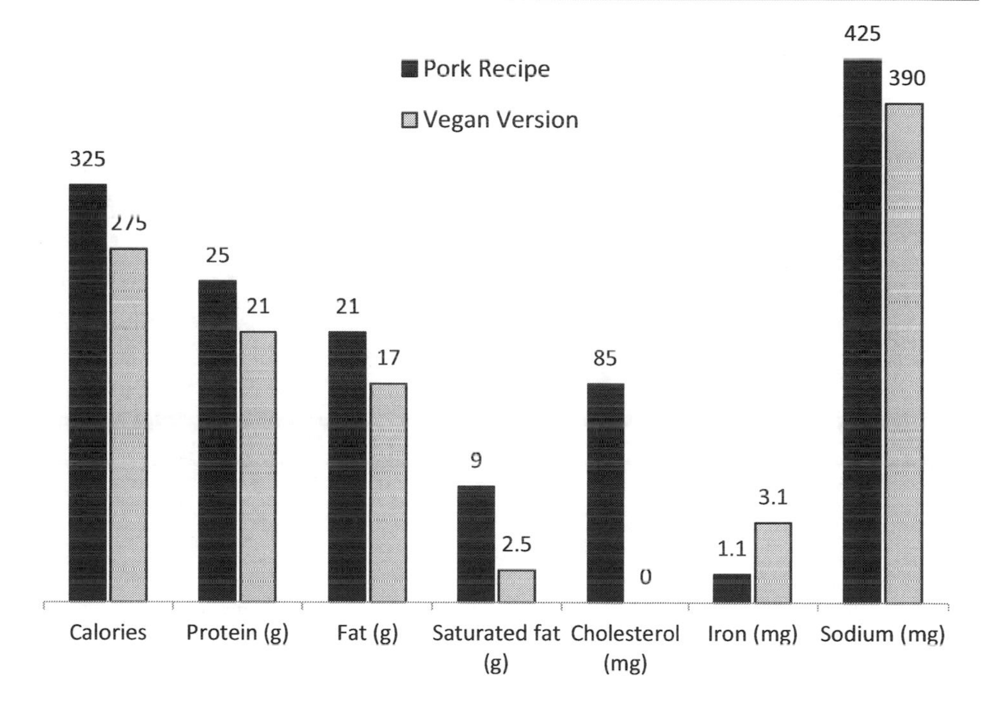

Baked Chops in Orange Sauce (6 servings)

Pork Recipe	Vegan Version
6 pork chops, about 1.6 lb (720 grams) total	*2 packages (enough for 720 grams) of extra-firm organic tofu, cut into ¼-inch (½-cm) slices*
2 Tbsp butter	2 Tbsp (30 ml) canola oil
2 orange bell peppers, sliced	2 orange bell peppers, sliced
1 small onion, chopped	1 small onion, diced
1 Tbsp cornstarch	1 Tbsp (15 ml) cornstarch
1 cup orange juice	1 cup (250 ml) orange juice
1 tsp vegetable broth powder	1 tsp (5 ml) vegetable broth powder
1 tsp pepper	1 tsp (5 ml) black pepper
1 tsp parsley flakes	1 tsp parsley flakes
1 tsp thyme	1 tsp (5 ml) thyme

Instructions:

1. Preheat oven to 350 °F (180 °C). Heat oil in a skillet at medium-high heat and brown the tofu slices, about 5 minutes per side. Remove from skillet and place in baking dish.
2. In the same skillet, add the bell peppers and onions. Sauté until onions are soft. In a bowl, combine the remaining ingredients and mix well.
3. Pour the liquid mixture back into the skillet at medium heat, and cook until sauce thickens. Remove from heat.
4. Pour the thickened orange sauce over the tofu slabs in the baking dish, and bake for 30 minutes.
5. Serve with brown basmati rice.

FACT: Pigs wiggle their tails when happy, just as dogs wag their tails. Pigs' social structure is akin to that of elephants. For more information, please visit the following site:
www.estherthewonderpig.com
Bacon, ham, and pork ribs come from piglets killed at the age of 6 months, whereas a pig's natural life span is about 12 years.

Environmental Savings

	Pork Recipe 6 chops @ 4oz (720 grams)	**Vegan Version** 1.6 lbs tofu (720 grams)	**Total Savings** (Pork minus Vegan)	**Savings Per Serving** (Pork minus Vegan)
H_2O	4,311 liters (1,139 US gal)	412 liters (109 US gal)	3,899 liters (1,030 US gal)	**650 liters (172 US gal)**
M	6.2 kg (13.6 lbs)	0	6.2 kg (13.6 lbs)	**1 kg (2.2 lbs)**
GHG	8.7 kg CO_{2e}	1.44 kg CO_{2e}	7.26 kg CO_{2e}	**1.2 kg CO_{2e} <=> (Driving 4.6 km or 2.9 miles)**

H_2O = Water Footprint, M =Manure produced, GHG = Greenhouse Gas

Nutritional Benefits per Serving

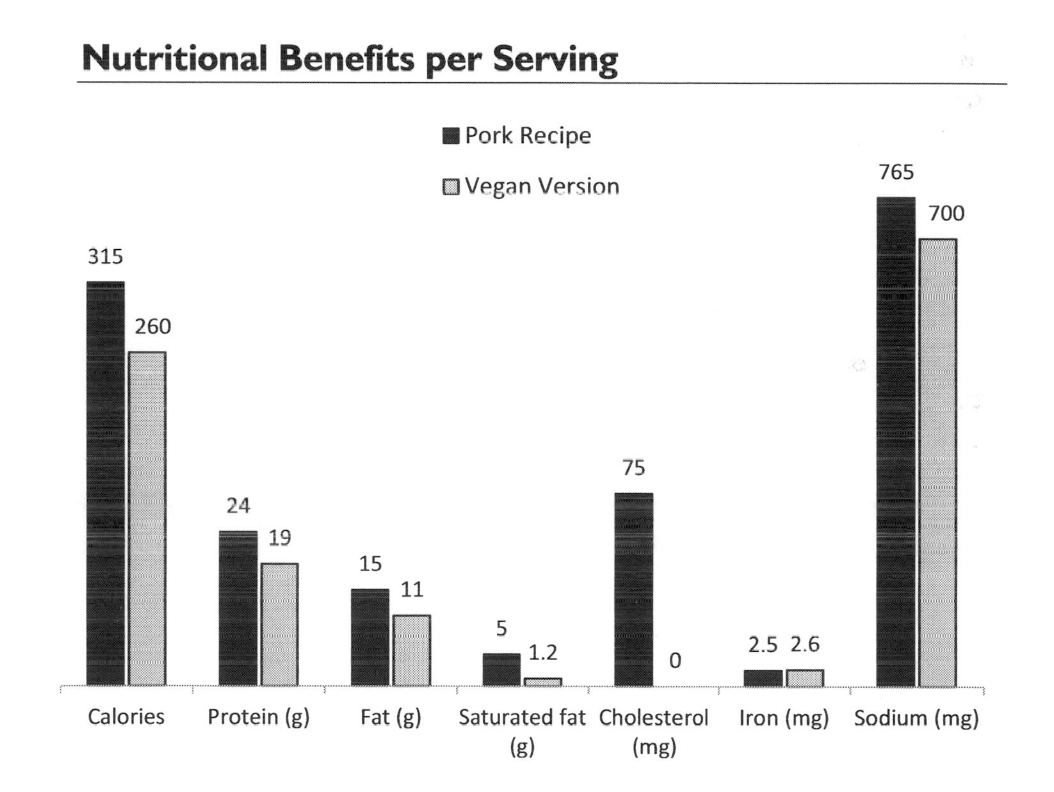

Simple Cabbage Rolls (8 rolls, 4 servings)

Pork Recipe	Vegan Version
1 lb (450 grams) ground pork	*1 lb (450 grams) firm or extra-firm organic tofu, crumbled*
1 Tbsp extra light olive oil	1 Tbsp (15 ml) extra light olive oil
1 small onion, chopped	1 small onion, chopped
2 cloves garlic, minced	2 cloves garlic, minced
1 medium green cabbage	1 medium green cabbage
½ tsp salt	½ tsp (3 ml) salt
1 tsp pepper	1 tsp (5 ml) pepper
3 Tbsp fresh parsley, minced	3 Tbsp (45 ml) fresh parsley, minced
1 cup cooked brown rice	1 cup (250 ml) cooked brown rice or quinoa
1 can tomato sauce	1 can tomato sauce
Pasta sauce as needed	Pasta sauce as needed
Louisiana hot sauce (optional)	Louisiana hot sauce (optional)

Instructions:

1. Blanch the head of cabbage in boiling water. Let cool, then separate into leaves (about 8).
2. In a large bowl, combine the crumbled tofu, onion, garlic, salt, pepper, parsley, rice or quinoa, oil, and tomato sauce. Mix together.
3. Preheat oven to 350 °F (180 °C). Spoon some filling onto the center of each cabbage leaf. Make a parcel by folding the stem end over first, then the sides. Roll them up as tightly as possible. Place in a casserole dish.
4. Repeat until all the cabbage leaves have been filled. Pour pasta sauce over the cabbage rolls and bake, covered, in the oven for ½ hour.
5. Serve with a dash of Louisiana hot sauce (if desired) and vegan sour cream.

FACT: Sadly, these intelligent and affectionate creatures are reduced to machines for yielding flesh and ribs for human consumption. If North Americans do not eat dogs, why eat the more intelligent and just as affectionate pig?

Environmental Savings

	Pork Recipe 1lb (450 grams)	Vegan Version Tofu 1 lb (450 grams)	Total Savings (Pork minus Vegan)	Savings Per Serving (Pork minus Vegan)
H_2O	2,695 liters (712 US gal)	257 liters (68 US gal)	2,438 liters (644 US gal)	**610 liters (161 US gal)**
M	3.9 kg (8.6 lbs)	0	3.9 kg (8.6 lbs)	**1 kg (2.2 lbs)**
GHG	5.45 kg CO_{2e}	.9 kg CO_{2e}	4.55 kg CO_{2e}	**1.1 kg CO_{2e} (Driving 4.2 km or 2.6 miles)**

H_2O = Water Footprint, M =Manure produced, GHG = Greenhouse Gas

Nutritional Benefits per Serving

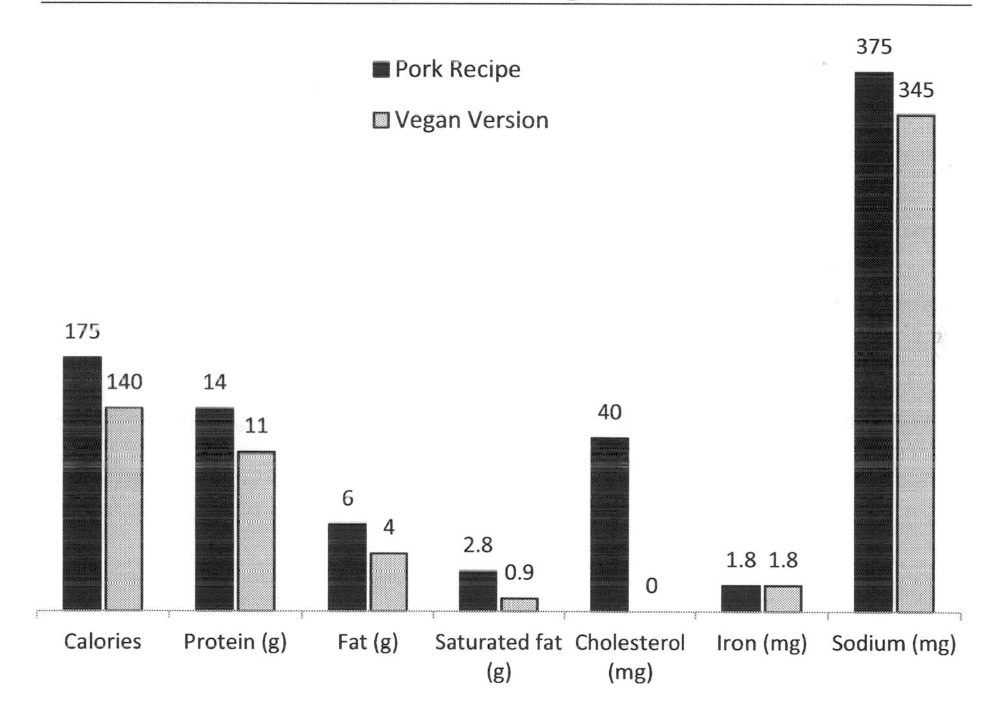

SUMMARY

ENVIRONMENTAL SAVINGS

(Pork minus Vegan)

Environmental Parameter per serving	Szechwan Ground Tofu with Green Beans	Baked Chops in Orange Sauce	Simple Cabbage Rolls
Water saved	610 liters (161 US gal)	650 liters (172 US gal)	610 liters (161 US gal)
Manure saved	1 kg (2.2 lbs)	1 kg (2.2 lbs)	1 kg (2.2 lbs)
GHG saved, in kg CO_{2e}	1.1 ⇔ driving 4.2 km (2.6 miles)	1.2 ⇔ driving 4.6 km (2.9 miles)	1.1 ⇔ driving 4.2 km (2.6 miles)

CONCLUSIONS

Environmental Savings

i. Although the values for all three environmental parameters are lower for pork than for beef, pork still is environmentally costly. The substitution of tofu for ground pork or pork chops resulted in water footprint savings ranging from about 610 to 650 liters (161 to 172 US gallons) per serving.

ii. Similarly, preparing these vegan versions spared the environment the production of 1 kg, or 2.2 lb, of manure per serving.

iii. Greenhouse gas emissions for the tofu recipes resulted in savings of 1.2 kg of CO_{2e}, which is equivalent to the emissions caused by driving 4.6 km (just under 3 miles).

iv. Since marinating, basting, and cooking tofu yields a taste and texture similar to those of pork, while reducing harmful environmental and health consequences, it makes sense to do so. This choice also would save large numbers of sentient animals from lives of pain, misery, and death.

SUMMARY

NUTRITIONAL BENEFITS

(Pork minus Vegan)

Nutritional Parameter per serving	Szechwan Ground Tofu with Green Beans	Baked Chops In Orange Sauce	SImple Cabbage Rolls
Calories	50 fewer	55 fewer	35 fewer
Protein	4 g less	5 g less	3 g less
Fat	4 g less	4 g less	2 g less
Saturated fat	6.5 g less	3.8 g less	1.9 g less
Cholesterol	85 mg eliminated	75 mg eliminated	40 mg eliminated
Iron	2 mg more	0.1 mg more	0 (identical)
Sodium	35 mg less	65 mg less	30 mg less

CONCLUSIONS

Nutritional Benefits

i. The replacement of pork with tofu creates appreciable, if less dramatic, savings in fat and in saturated fat than those obtained through replacing beef. Yet, the saturated fat savings realized by the Baked Chops recipe is great enough to substantially reduce cardiovascular risk.

ii. While the protein content of the vegan versions is slightly less than that of traditional pork recipes, the iron content of the vegan versions is identical or even higher. You can enjoy flavor, texture, and good nutrition without compromising your health and without killing animals.

iii. During the advertising campaign for "the other white meat," intended to highlight pork's allegedly low fat content, the per capita consumption of pork rose only 7.5 percent. By the campaign's end in 2011, per capita consumption was lower than it had been at the time of the campaign's launching. The per capita intake of this truly red meat is more than 20 percent lower than it was in 1980, suggesting that consumers are waking up to the dangers inherent in eating pork.

Chapter 5

How Veganizing Chicken Recipes Improves Environmental and Nutritional Parameters: *An Analysis*

Is Chicken a Health Food?

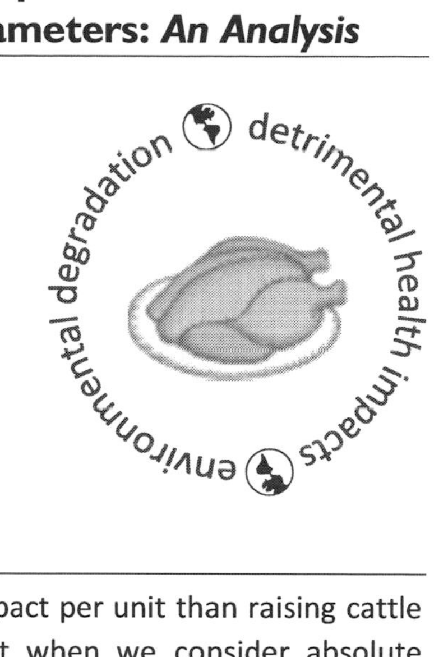

ENVIRONMENT

Raising chickens has less environmental impact per unit than raising cattle or pigs. However, these savings are lost when we consider absolute numbers.

Although the water footprint per kg for chicken is similar to that for beans, the water footprints per unit of protein are quite different (Chapter 1, Figure 4). At 34 liters of water per gram of protein, chicken has a higher water footprint than the 19 liters per gram of protein for beans. Thus, it requires almost twice as much water to produce protein from chickens than it does protein from beans.

Industrial chicken barns can hold between 5,000 and 170,000 chickens. In 2013, Americans ate 8.6 billion chickens. This figure constitutes 95% of the 9 billion commonly farmed animals slaughtered for food annually. Maryland and Delaware together produce over 500 million chickens a year, resulting in an estimated 40 million cubic feet of manure. This amount would entirely fill the US Capitol dome almost once a week, every week of the year. This volume of waste presents detrimental consequences for Chesapeake Bay, one of America's most threatened bodies of water.

In Maryland and North Carolina, runoff from chicken and hog factory farms is believed to be partly responsible for the *Pfiesteria piscicida* outbreaks that have resulted in the deaths of millions of fish. This microorganism also is suspected of having caused the cognitive problems and skin irritations suffered by the local populace. The growth in size of industrial poultry operations is such that 10 companies currently produce over 90 percent of poultry in the US. That translates to a lot of chicken manure, densely concentrated in a few areas of that country.

Although the GHG emissions (Chapter 1, Figure 1) from chicken (6.9 kg of CO_{2e} per kg of meat) represent an improvement over those from red meat (27 kg of CO_{2e} per kg of beef; and 12.1 kg of CO_{2e} per kg of pork) and from turkey (10.9 kg CO_{2e}), nonetheless they are substantially higher than the emissions for plant foods high in protein such as beans and tofu (2 kg of CO_{2e}), or lentils (0.9 kg of CO_{2e}).

HEALTH

The government and mainstream health authorities recommend chicken as a health food, since it generally contains less total fat and saturated fat than red meat. Yet many consumers find that their weight, blood cholesterol, or blood pressure remains high even after they have replaced red meat in their diets with chicken. Furthermore, while eating a chicken breast is better than devouring a sirloin steak, neither can be regarded as health food; nor is either necessary to human nutrition. In fact, both are potentially injurious. When skeletal muscle (i.e., animal flesh including beef, pork, poultry, and fish) is cooked at high temperatures (above 300 °F) as in grilling, deep-frying, and broiling, or for a long time, heterocyclic amines (HCAs), are formed. As noted in the National Cancer Institute Fact Sheet, examples include grilled, barbecued, or well done chicken and steak, all of which contain high concentrations of HCAs. These chemical compounds are capable of altering DNA synthesis.

The government and mainstream health authorities recommend chicken as a health food, since it generally contains less total fat and saturated fat than red meat.

Yet many consumers find that their weight, blood cholesterol, or blood pressure remains high even after they have replaced red meat in their diets with chicken.

The National Cancer Institute's NIH-AARP (National Institute of Health) Diet and Health Study is the largest study of its kind, involving half a million Americans aged 50 – 71. It concluded (among other findings) that a diet including animal fat is linked to a higher risk of pancreatic cancer. This is one of the most aggressive forms of cancer; the five-year survival rate is a mere three percent, with a median survival time of less than six months. Since FDA-approved therapies engender severe side effects and may not result in remission, obviously prevention is the best way to beat pancreatic cancer.

Governmental dietary recommendations typically seek to limit foods that pose nutritional risks, rather than to encourage the consumption of foods that offer protection against disease. For instance, neither pork nor chicken breast provides benefits other than protein, B vitamins, and minerals. They supply no fiber, a crucial dietary element for lowering cholesterol. They supply only insignificant amounts of potassium and magnesium, which are important minerals for lowering blood pressure. Worse, a large number of chicken products in the retail market contain added salt water (aka "seasoning solution" or "chicken broth").

...a diet including animal fat is linked to a higher risk of pancreatic cancer. This is one of the most aggressive forms of cancer; the five-year survival rate is a mere three percent, with a median survival time of less than six months.

Saturated fat and cholesterol are not the only factors in arterial clogging. In 2013, a study by RA Koeth, et. al., at the Cleveland Clinic found that people who consumed the most carnitine, a substance present in animal flesh, increased their risk for heart disease by producing more artery-clogging metabolites. That same year, a study by WHW Tang, et. al., published in the *New England Journal of Medicine*, found that people with the highest levels of a choline by-product (choline being abundant in animal products) were 2.5 times more likely to suffer an adverse cardiovascular event than those with the lowest levels.

Many human cancer cells rely upon the amino acid methionine, without which they cannot survive. That particular amino acid comes principally from food. Dr. Michael Greger of NutritionFacts.org states that, to combat the majority of lethal metastatic cancers that are not responsive to chemotherapy, dietary methionine restriction may prove beneficial. However, fish and chicken (usually considered healthier choices) contain the highest levels of methionine, followed by eggs, red meat (beef and pork), and dairy products. Hence, the strategy of methionine restriction can best be achieved by adopting a plant-based diet.

Many human cancer cells rely upon the amino acid methionine, without which they cannot survive. However, fish and chicken (usually considered healthier choices) contain the highest levels of methionine, followed by eggs, red meat (beef and pork), and dairy products. Hence, the strategy of methionine restriction can best be achieved by adopting a plant-based diet.

When we replace chicken with organic tofu or beans, we are enriching our diet with fiber, phytochemicals, and significant mineral content, in addition to protein. At the same time, we are enhancing our ability to prevent or to treat obesity, high blood pressure, and heart disease.

A further reason for health concerns regarding chicken is the high prevalence of contamination from bacteria such as *E. coli*, *salmonella*, and *campylobacter*, all of which originate in the intestines of animals and humans. In one study, up to 80 percent of all retail chicken randomly purchased was found to be contaminated with one or more pathogenic bacteria strains. Since plant foods have no intestines, any contamination of plant foods can only be a direct consequence of improper human handling or of problems with processing equipment or with farm runoff. Unlike the dangers *inherent* in red meat and chicken, against which there can be no completely adequate safeguards, the potential and lesser dangers arising from plant food can be eliminated through proper training and enforcement.

"The indifference, callousness and contempt that so many people exhibit toward animals is evil first because it results in great suffering in animals, and second because it results in an incalculably great impoverishment of the human spirit. All education should be directed toward the refinement of the individual's sensibilities in relation not only to one's fellow humans everywhere, but to all things whatsoever."

~Ashley Montagu, Humanist of the Year, 1995

Chicken and Apricot Salad (3 servings)

Chicken Recipe	Vegan Version
1 stalk celery, chopped	1 stalk celery, chopped
½ cup dried apricots, chopped into tidbits	½ cup (125 ml) dried apricots, chopped into tidbits
11 oz chopped, cooked chicken meat	*11 oz (about 310 grams) extra-firm organic tofu, cubed*
¼ cup light mayonnaise	¼ cup (60 ml) vegan mayonnaise
¼ cup hummus	¼ cup (60 ml) hummus
2 tsp lemon juice	2 tsp (10 ml) lemon juice
½ teaspoon ground black pepper	½ (3 ml) teaspoon ground black pepper

Instructions:

1. Chop the celery and dried apricots. Chop the tofu into bite-sized cubes.
2. In a medium bowl, mix together the mayonnaise, hummus, lemon juice, and pepper. Add the celery, apricots, and tofu, and combine until well coated.

FACT: In a review of 20 years of research on chicken intelligence, Bristol University's professor of animal welfare, Christine Nicol, found that newly born chicks can count up to five and exhibit intelligent behavior within a few hours of hatching. She concluded that they exhibit skills that only children four or older can perform. Her paper, The Intelligent Hen, indicates that chickens are born with an innate understanding of structural engineering through experiments that displayed their interest in diagrams of objects that can be built, rather than those that defy the law of physics.

Chickens also have empathy; they can plan ahead and exhibit self-control. This was demonstrated by birds deciphering that if they waited longer to start eating, they would have longer access to food, a behavior attributable to children only after age four.

Environmental Savings

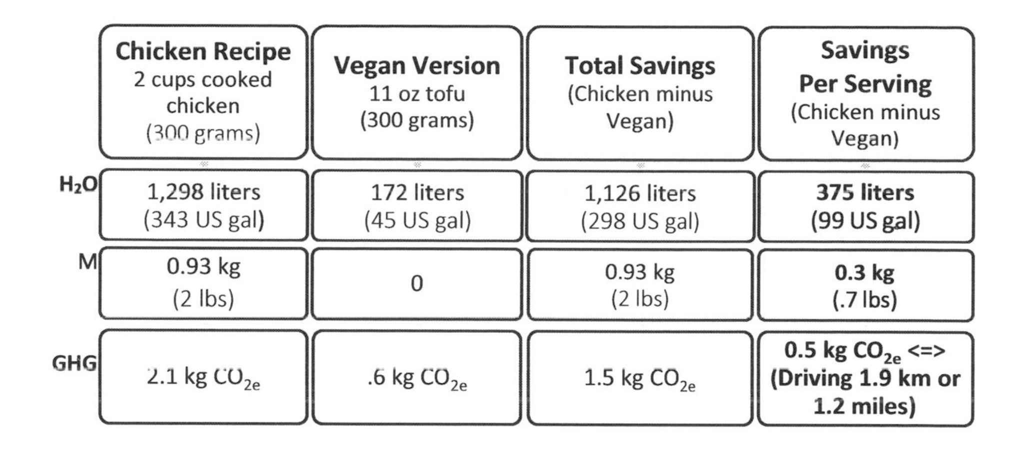

	Chicken Recipe 2 cups cooked chicken (300 grams)	Vegan Version 11 oz tofu (300 grams)	Total Savings (Chicken minus Vegan)	Savings Per Serving (Chicken minus Vegan)
H_2O	1,298 liters (343 US gal)	172 liters (45 US gal)	1,126 liters (298 US gal)	**375 liters (99 US gal)**
M	0.93 kg (2 lbs)	0	0.93 kg (2 lbs)	**0.3 kg (.7 lbs)**
GHG	2.1 kg CO_{2e}	.6 kg CO_{2e}	1.5 kg CO_{2e}	**0.5 kg CO_{2e} <=> (Driving 1.9 km or 1.2 miles)**

H_2O = Water Footprint, M =Manure produced, GHG = Greenhouse Gas

Nutritional Benefits per Serving

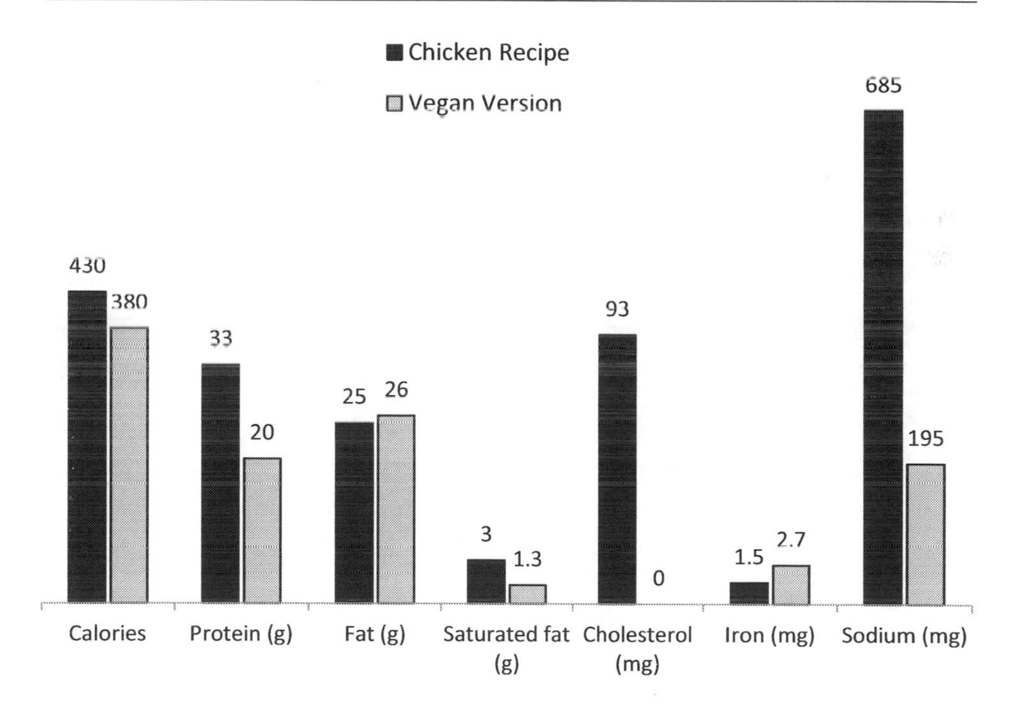

Ginger-Sesame Chicken (4 servings)

Chicken Recipe	Vegan Version
1 lb (450 grams) chicken breasts	*1 lb (450 g) extra-firm organic tofu, cut into 4 equal slabs*
2 tsp shortening	2 tsp (10 ml) canola oil
½ cup carrots, julienned	½ cup (125 ml) carrots, julienned
¾ cup light sesame-ginger salad dressing	¾ cup light sesame-ginger salad dressing
1 Tbsp fresh ginger	1 Tbsp (15 ml) fresh ginger, chopped
3 Tbsp sesame seeds	3 Tbsp (45 ml) sesame seeds

Instructions:

1. Preheat oven to 350 °F (180 °C). Cut the tofu into 4 equal slabs and place into baking dish.
2. Heat the oil in a skillet and pan-fry the julienned carrots for 5 – 10 minutes, until softened. Reduce heat and add remaining ingredients except sesame seeds. Heat for a few minutes.
3. Pour the marinade over the tofu. Sprinkle with sesame seeds and bake for 30 minutes.

FACT: Chickens have family units and social hierarchies. They enjoy sun- and dust-bathing activities, and diligently care for their hatchlings. They are social creatures with individual personalities. Researchers from the Imperial College in London, England discovered that the areas of birds' long-term memory and problem solving are wired similarly to comparable areas of the human brain. Chickens can recognize up to 100 individuals and have up to 30 vocalizations to communicate with one another.

Environmental Savings

	Chicken Recipe 1 lb chicken (450 grams)	Vegan Version 1 lb tofu (450 grams)	Total Savings (Chicken minus Vegan)	Savings Per Serving (Chicken minus Vegan)
H_2O	1,946 liters (514 US gal)	257 liters (68 US gal)	1,689 liters (446 US gal)	**422 liters (112 US gal)**
M	1.4 kg (3.1 lbs)	0	1.4 kg (3.1 lbs)	**0.35 kg (.8 lbs)**
GHG	3.1 kg CO_{2e}	.9 kg CO_{2e}	2.2 kg CO_{2e}	**0.55 kg CO_{2e} <=> (Driving 2.1 km or 1.3 miles)**

H_2O = Water Footprint, M =Manure produced, GHG = Greenhouse Gas

Nutritional Benefits per Serving

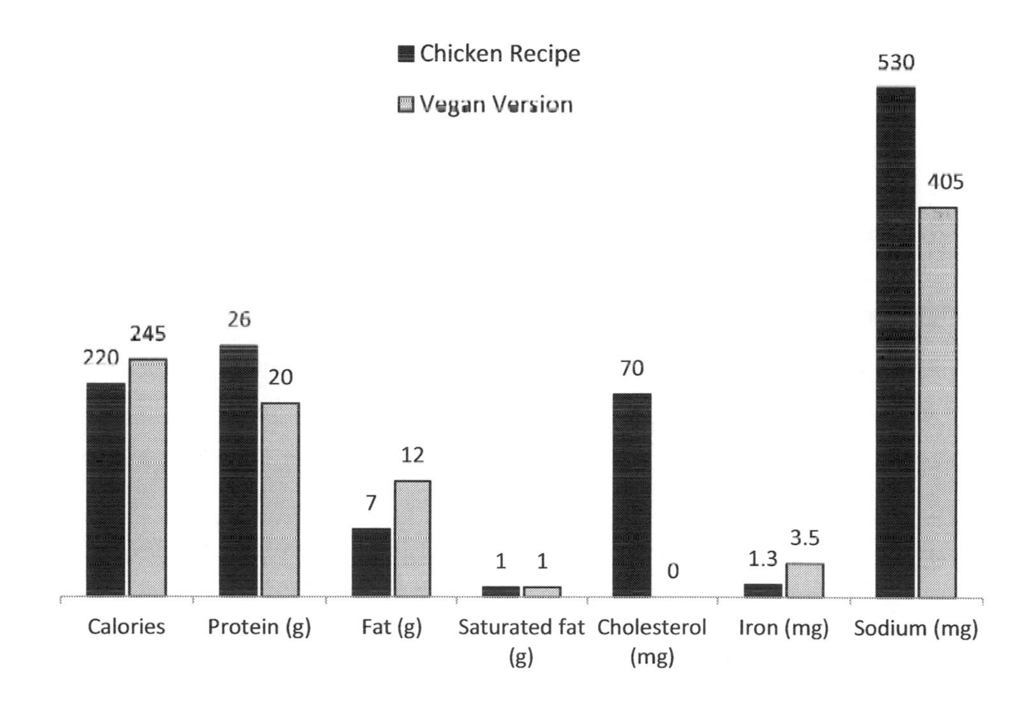

Italian-Style Stew (4 servings)

Traditional Recipe	Vegan Version
1 Tbsp olive oil	1 Tbsp (15 ml) canola oil
1 small onion, diced	1 small onion. diced
1 medium yellow pepper, diced	1 medium yellow pepper, diced
1 cup chopped zucchini	1 cup (250 ml) chopped zucchini
2 Tbsp chopped fresh parsley	2 Tbsp (30 ml) chopped fresh parsley
1 Tbsp chopped fresh basil	1 Tbsp (15 ml) chopped fresh basil
1 lb (450 grams) chicken breasts, diced	*3 cups (750 ml) cooked navy beans*
1 15-oz can Italian diced tomatoes	1 15-oz can (400 ml) Italian diced tomatoes

Instructions:

1. In a large pot on medium heat, sauté onion, yellow pepper, and zucchini in oil until soft. Add parsley, basil, beans, and tomatoes.
2. Reduce to a simmer and cook for another 20 minutes.

FACT: Each of the 623 million chickens in Canada and the 8.6 billion chickens in the US that are bred as broilers grow at six times their natural rate. A chicken's normal lifespan is about 10 years, but the chicken body parts that people eat come from chicks that are only 7 weeks old.

During their short lives, they suffer from deformities because of having been bred to grow so quickly that their legs cannot support their own weight. At 5 weeks old, they usually are too heavy to walk or to even move around. This weight gain is equivalent to a 2-year-old human baby weighing 350 pounds!

What if this practice were to be extended to other animals, such as dogs and cats? Would it then enable us to recognize the travesty we impose on chickens for the sake of our palates?

Environmental Savings

	Chicken Recipe 1 lb chicken breast (450 grams)	**Vegan Version** 3 cups cooked navy beans (225 grams)	**Total Savings** (Chicken minus Vegan)	**Savings Per Serving** (Chicken minus Vegan)
H_2O	1,946 liters (514 US gal)	912 liters (241 US gal)	1,034 liters (273 US gal)	**256 liters (68 US gal)**
M	1.4 kg (3.1 lbs)	0	1.4 kg (3.1 lbs)	**0.35 kg (.8 lbs)**
GHG	3.1 kg CO_{2e}	.45 kg CO_{2e}	2.65 kg CO_{2e}	**0.67 kg CO_{2e} <=> (Driving 2.6 km or 1.6 miles)**

1 cup dry beans weighs 8 oz (225 grams), yields 3 cups cooked

H_2O = Water Footprint, M =Manure produced, GHG = Greenhouse Gas

Nutritional Benefits per Serving

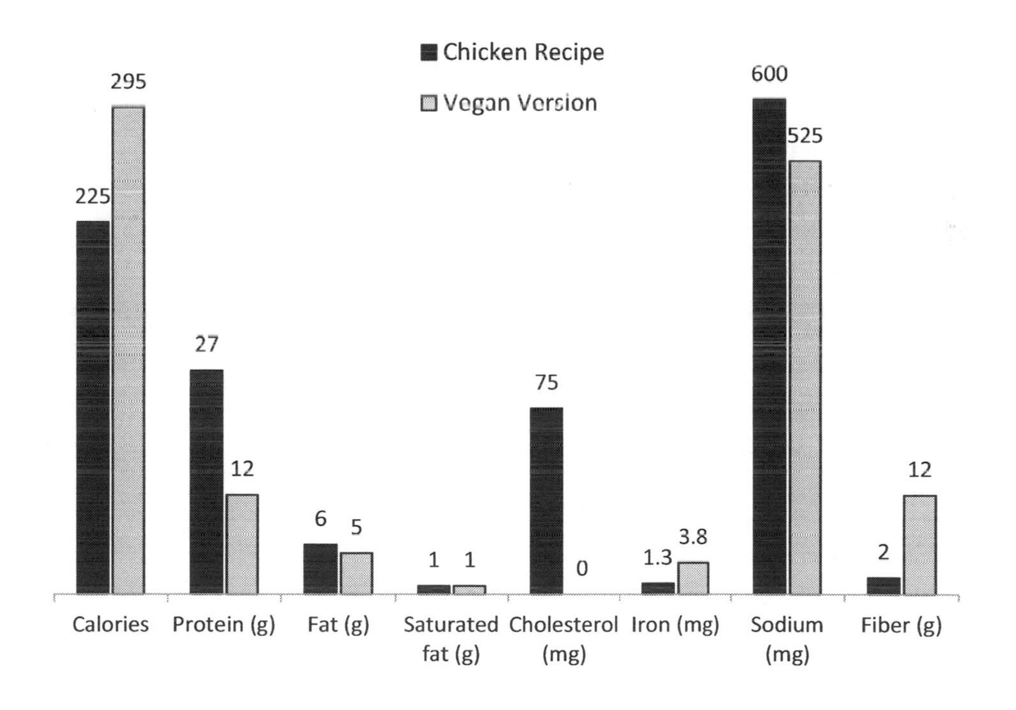

SUMMARY

Environmental Savings

(Chicken minus Vegan)

Environmental Parameter (per serving)	Chicken and Apricot Salad	Ginger-Sesame Chicken	Italian-Style Stew
Water saved	375 liters (99 US gal)	422 liters (112 US gal)	256 liters (68 US gal)
Manure saved	0.3 kg (0.7 lbs)	0.35 kg (0.8 lbs)	0.35 kg (0.8 lbs)
GHG saved, in kg of CO_{2e}	0.5 ⇔ driving 1.9 km (1.2 miles)	0.55 ⇔ driving 2.1 km (1.3 miles)	0.67 ⇔ driving 2.6 km (1.6 miles)

CONCLUSIONS

Environmental Savings

i. The weight of raw chicken includes water and fat. In contrast, uncooked beans have negligible amounts of water and fat. Nutritionally, beans are superior in that they have far less total and saturated fat than chicken meat. The water savings in the sample recipes show a range of 256 to 422 liters (68 to 112 US gallons) per serving, a sizeable saving relative to personal water conservation efforts.

ii. The manure savings provided by these recipes approximate 0.35 kg, or 0.8 lbs, per serving. Because chickens are smaller than pigs or cattle, they produce less waste per animal. However, the greater number of chickens (as compared to other animals) consumed by humans leads to an extremely large absolute total amount of chicken manure.

iii. GHG emissions for chicken are roughly 7 kg per kg of chicken meat. This figure still is substantially higher than that for plant proteins. For instance, broccoli, tofu, and beans each have GHG emissions of only 2 kg per kg of food. Chicken, therefore, represents a threefold increase in GHGs. In our illustrative recipes, GHG reductions range from 0.5 to 0.67 kg of CO_{2e} per serving. That translates to driving between 1.9 km (1.2 miles) and 2.6 km (1.6 miles).

iv. Although the manure and GHG savings obtained in substituting beans or tofu for chicken are less than those for beef and pork, saving 256 liters (68 US gal) of water per serving of Italian-Style Stew is noteworthy.

SUMMARY

Nutritional Benefits
(Chicken minus vegan)

Nutritional Parameter (per serving)	Chicken and Apricot Salad	Ginger-Sesame Chicken	Italian-Style Stew
Calories	50 fewer	25 more	70 more
Protein	13 g less	6 g less	15 g less
Fat	1 g more	5 g more	1 g less
Saturated fat	1.7 g less	0 (identical)	0 (identical)
Cholesterol	93 mg eliminated	70 mg eliminated	75 mg eliminated
Iron	1.2 mg more	2.2 g more	2.5 mg more
Sodium	490 mg less	125 mg less	75 mg less
Fiber	–	–	10 g more

CONCLUSIONS

Nutritional Benefits

i. In the Chicken and Apricot Salad recipe, there is a huge sodium reduction by replacing chicken with tofu. In the Ginger-Sesame Chicken, the tofu effectively replaced the saturated fat with healthier, unsaturated fats. In the Italian-Style Stew, the use of beans instead of chicken added five times more fiber, and more than doubled the iron content.

ii. These analyses were based on comparing the leanest chicken parts to tofu or beans. If chicken dishes were prepared using the higher-fat body parts, such as thighs, legs, or wings the nutritional advantages of the vegan versions of these recipes would be even greater.

iii. Nowadays, many meat substitutes have similar tastes and textures to chicken meat. These plant-based alternatives can be substituted directly in traditional chicken recipes.

Chapter 6

How Veganizing Dairy and Egg Recipes Improves Environmental and Nutritional Parameters: *An Analysis*

Is It Natural for Humans to Drink Cows' Milk?

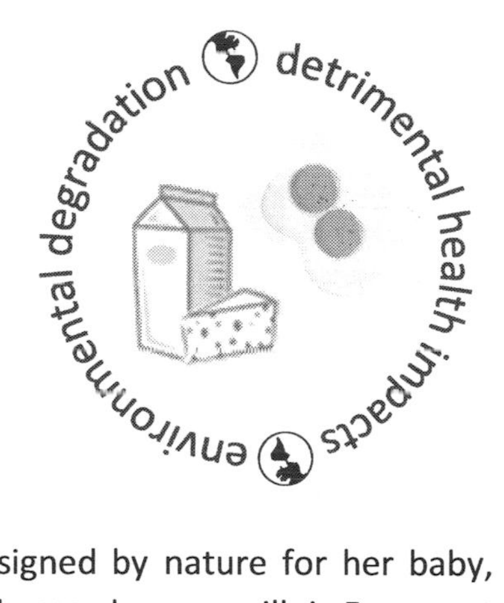

All mammals, including humans, produce milk to nourish their babies. But why do we, as adults, consume the milk made for infants of another species? When considered in this light, doesn't our practice of drinking cows' milk seem unnatural? Just as a woman's breast milk is designed by nature for her baby, cows' milk is nature's perfect food for calves. Whereas human milk is 5 percent protein, cows' milk has a 15-percent protein content, which enables a calf to double its weight in 45 days. Yet it takes human babies six months to double their weight. Thus, cows' milk is too high in protein to be appropriate for human consumption—particularly for adults, who have stopped growing. Does our consumption of cows' milk represent an attempt to compensate for an imagined nutritional requirement?

Over 60 percent of adults worldwide lose the ability to digest lactose after childhood. Lactose intolerance, which varies according to ethnicity, ranges from 75 percent for Africans to 95 percent for Asians, and to nearly 100 percent for Native Americans. On the other hand, only 5 percent of northwestern Europeans and of Scandinavians have lactose intolerance. In fact, scientists maintain that the continued ability to digest lactose is abnormal and represents an aberration rather than the norm, as viewed by many people.

In the US, allergy to cows' milk affects 2.5 percent of children under the age of three, while soy milk allergy affects only 0.3 percent of children. Dairy products are touted for their calcium content; however, only one-third of this calcium actually is absorbed. In contrast, the human body absorbs 50 percent of the calcium contained in plant sources such as collards, kale, and bok choy.

ENVIRONMENT

Since roughly 10 cups of cows' milk are required to make 1 cup of cheese, its environmental costs are very high (Chapter 1, Figure 2). For instance, cheese has a water footprint of 5,000 liters per kg of cheese, which places it between pork and chicken in this regard. Additionally, cheese ranks high in manure production (Chapter 1, Figure 5) at 37 kg of manure per kg of cheese. It also is the third highest in terms of GHG emissions, after lamb and beef. The many plant-based cheese alternatives now in the marketplace may not offer significantly less fat content, but they have the advantages of being less damaging to the environment and do not require the use of animals.

...cheese has a water footprint of 5,000 liters per kg of cheese, which places it between pork and chicken in this regard. Additionally, cheese ranks high in manure production at 37 kg of manure per kg of cheese. It also is the third highest in terms of GHG emissions, after lamb and beef.

The recent trend toward Greek-style yogurt has precipitated an unexpected environmental problem: the disposal of acid whey. Whereas traditional yogurt requires a ratio of 1 to 1.3 milk portion to produce 1 yogurt portion, Greek-style yogurt demands a ratio of up to 4:1. This results in an accumulation of acid whey, in addition to a water footprint that is 3 to 4 times that of milk. In traditional yogurt, whey is mixed in and kept, but Greek-style yogurt filters out the acid whey so as to concentrate the protein and to thicken the texture. The USDA is now exploring new technology to address the issue of acid whey disposal. In 2012, its Agricultural Research Service partnered with the food innovation firm Jones Laffin to develop a technology to extract whey protein and lactose from acid whey. Used only in limited amounts in animal feed, acid whey is a pollutant that, if released into waterways, will result in massive fish kills by depleting oxygen.

Non-dairy alternatives on the market include not only organic soy milk but also oat, rice, coconut, hemp, and almond milk. However, nut milks such as almond milk have a comparable water footprint to dairy milk. In contrast, soy milk's water footprint is less than half that of dairy. Additionally, soy milk`s calcium and protein content make it an excellent substitute for dairy. The occasional use of other non-dairy alternatives including nut milk would offer variety while not imposing a significant environmental toll.

HEALTH

In an interview with Cherrill Hicks of *The Daily Telegraph* in June of 2014, Dr. Jane Plant states, "The growth factors and hormones cows' milk contains are not just risky for breast cancer, but also [for] other hormone-related cancers, of the prostate, testicles and ovary." A highly regarded professor at Imperial College in London, England who specializes in environmental carcinogens, Dr. Plant is also a six-time cancer survivor. Her message is that a diet excluding dairy will deprive cancer cells of the conditions they need to grow. She reveals that cows' milk contains 35 different hormones and 11 growth factors. One in particular, IGF-1, is strongly linked to the development of many cancers. "This means that a vegan diet is lower in cancer-promoting molecules and higher in the binding proteins that reduce the action of these molecules."

Scientists at Kaiser Permanente research center in California found that women with breast cancer who ate just one portion of yogurt, cheese, or ice cream a day were 50 percent more likely to die from the disease within 12 years. They emphasized that most dairy products in Britain and in the US come from pregnant cows and thus are rich in estrogen, which encourages tumor growth.

...women with breast cancer who ate just one portion of yogurt, cheese, or ice cream a day were 50 percent more likely to die from the disease within 12 years.

The American College of Nutrition latest recommendations include limiting or avoiding dairy products so as to reduce prostate cancer risk. Joseph Gonzales, the lead author and Registered Dietitian at the University of Texas, MD Anderson Cancer Center indicated, "Plants are rich in protective compounds and help consumers avoid the cancer-causing substances found in animal products. Plant-based foods also have a slimming effect, which reduces risk for all forms of cancer in the long run." The study found that eating just 35 grams of dairy protein per day increases prostate cancer risk by 32

The American College of Nutrition latest recommendations include limiting or avoiding dairy products so as to reduce prostate cancer risk.

The study found that eating just 35 grams of dairy protein per day increases prostate cancer risk by 32 percent. And drinking two glasses of milk will increase that risk by 60 percent.

percent. That's the equivalent of 1.5 cups of cottage cheese. And drinking two glasses of milk will increase that risk by 60 percent.

When we replace dairy with non-dairy equivalents, we may not eliminate fat and sodium content, but we do eliminate the hormones and growth factors, as well as the opioid compounds naturally occurring in cow's milk. In his book, *Breaking the Food Seduction*, Dr. Neal Barnard points out that these addictive compounds may serve to explain the high rates of cheese consumption. Over the last 40 years, cheese has played an increasing role in the American diet. The average American now eats three pounds of cheese per month. That translates into a significant intake of fat, saturated fat, and sodium. While vegan "cheese" products are not health foods, they do not exert any hormonal influences upon us and do not possess the addictive qualities typical of cheese.

Most plant-based milks have comparable calcium content to cows' milk, and are available in major supermarkets. Consumers are advised to select products containing less than 13 g sugar per cup. These products may be used, in identical quantities, to replace dairy milk in nearly any recipe.

A 2013 meta-analysis of 14 studies found that people who consumed the most eggs had a 19-percent increased risk of developing cardiovascular disease and a 68-percent increased risk for diabetes, compared to people eating the fewest eggs.

In the following analysis, dairy cheese is replaced with Daiya vegan "cheese" in a 1:1 ratio. Daiya is commonly available throughout the US and Canada, and is soy-free, nut-free, and gluten-free. However, Daiya is not a whole food. Those watching their sodium and fat content may wish to reduce the amount of replaced Daiya "cheese" by about 15 percent.

"A man can live and be healthy without killing animals for food; therefore, if he eats meat, he participates in taking animal life merely for the sake of his appetite. And to act so is immoral."

~ Leo Tolstoy

Simple Cheese Scones (6)

Dairy Recipe	Vegan Version
2 cups all-purpose flour	2 cups (500 ml) whole-wheat flour
2 tsp baking powder	3 tsp (15 ml) baking powder
½ tsp salt	—
½ tsp cream of tartar	½ tsp (3 ml) cream of tartar
1 Tbsp sugar	1 Tbsp (15 ml) sugar
½ cup shortening	½ cup (125 ml) vegan margarine
1½ cups grated cheddar cheese, about ½ lb	*1½ cups (375 ml) Daiya cheddar shreds*
1 large egg	*Flax Egg (optional)**
½ cup milk	*½ – ¾ cup (125 – 185 ml) soy milk or other plant-based milk*

**A "flax egg" [made of 1 Tbsp (15 ml) ground flax seed and 3 Tbsp (45 ml) water] may be added to simulate the consistency of a real egg.*

Instructions:

1. Preheat oven to 425 °F (220 °C).
2. Combine the flour, baking powder, cream of tartar, and sugar in a bowl. Cut the vegan margarine into the dry mixture until it resembles coarse crumbs. Fold in Daiya shreds.
3. If adding a "flax egg" (for texture): In a small, separate bowl, mix 3 Tbsp of water into 1 Tbsp ground flax seed, stirring continuously. Refrigerate for at least 15 minutes, until gelatinous. Then add "flax egg" to mixture.
4. To form dough, add soy milk or other plant-based milk to mixture. Do not knead the dough excessively. Place dough on cookie sheet and cut it into desired number of wedges. Bake for about 20 minutes, until it starts to brown. Remove promptly from oven, so that scones do not become dry.

FACT: About 50 percent of newly hatched chicks are males. All are either ground up live in a high-speed grinder (widely used in the US and Canada), gassed to death (common in the UK), or electrocuted, when less than 72 hours old. Both conventional and organic farms are supplied by hatcheries employing these practices. The egg industry kills 40 million male chicks in the UK each year, and 250 million male chicks in the US.

Environmental Savings

	Dairy Recipe 1.5 cups (6 oz) cheddar cheese (0.17 kg)	**Dairy Recipe** 1 egg (0.05 kg)	**Dairy Recipe** ½ cup milk (0.125 kg)	**Vegan Version** 1.5 cups (6 oz) Daiya cheddar shreds (0.17 kg)	**Vegan Version** "Flax egg" uses 1 Tbsp ground flax seed (7 g)	**Vegan Version** ½ - ¾ cup soy milk (0.19 kg)
H_2O	850 liters (225 US gal)	163 liters (43 US gal)	128 liters (34 US gal)	282 liters (75 US gal)	36 liters (10 US gal)	57 liters (15 US gal)
M	6.3 kg (13.9 lbs)	0.11 kg (0.2 lbs)	0.46 kg (1 lb)	0	0	0
GHG	2.3 kg CO_{2e}	0.24 kg CO_{2e}	0.24 kg CO_{2e}	1 kg CO_{2e}	0.016 kg CO_{2e}	0.19 kg CO_{2e}

	Total Savings (Dairy minus Vegan)	**Savings Per Scone** (Dairy minus Vegan)
H_2O	766 liters (202 US gal)	**120 liters (34 US gal)**
M	6.9 kg (15.1 lbs)	**1.15 kg (2.5 lbs)**
GHG	1.57 kg CO_{2e}	**0.26 kg CO_{2e} (Driving 1 km or 0.6 mile)**

1 lb eggs = 9 Grade A large eggs = 0.45 kg

Water Footprint, M =Manure produced, GHG = Greenhouse Gas

Nutritional Benefits per Serving

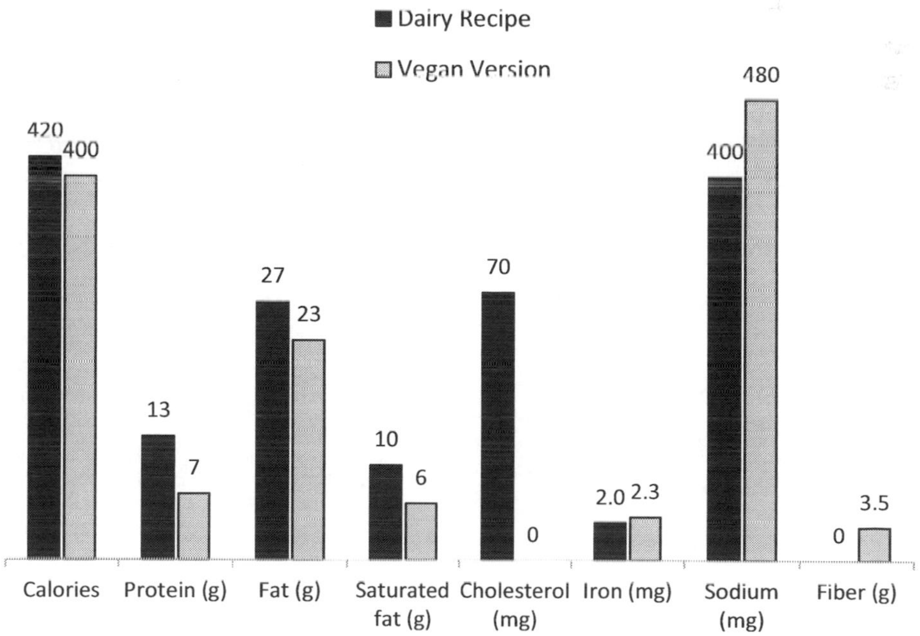

Mac and Cheese (6 servings)

Dairy Recipe	Vegan Version
2 cups whole-wheat elbow macaroni	2 cups (500 ml) whole-wheat elbow macaroni
½ tsp salt	—
½ tsp pepper	½ tsp (3 ml) pepper
1½ cups milk	*1½ cups (375 ml) soy milk*
1 Tbsp all-purpose flour	1 Tbsp (15 ml) whole-wheat flour
2 Tbsp margarine	2 Tbsp (30 ml) vegan margarine
1 cup shredded cheddar cheese	*1 cup (250 ml) Daiya cheddar shreds*
1 cup shredded mozzarella cheese	*1 cup (250 ml) Daiya mozzarella or pepper jack shreds*

Instructions:

1. Preheat oven to 350 °F (180 °C).
2. Cook elbow macaroni on stovetop according to package instructions, until *al dente* (about 8 – 10 minutes). Drain and place in baking dish.
3. In a saucepan, heat the vegan margarine, then add pepper.
4. Stir in the flour, and mix well. Pour in the soy milk, stirring constantly. Bring to gentle boil and then simmer for several minutes, until thickened.
5. Remove saucepan from heat and fold in Daiya cheddar and mozzarella shreds. Combine until melted. Pour mixture over cooked macaroni in baking dish.
6. Bake for about 20 – 25 minutes, until heated through.

FACT: Canadian and—to a lesser extent—American farms implement forced moulting, whereby food and water are withheld from laying hens for up to 14 days.

FACT: Cows and chickens are more intelligent than the dogs and cats cherished as pets, and certainly are equal in their ability to feel pain and in their desire to live.

Environmental Savings

	Dairy Recipe 2 cups (8 oz) shredded cheese (0.225 kg)	**Dairy Recipe** 1.5 cups milk (0.375 kg)	**Vegan Version** 2 cups (8 oz) Daiya shreds (0.225 kg)	**Vegan Version** 1.5 cups soy milk (0.375 kg)	**Total Savings** (Dairy minus Vegan)	**Savings Per Serving** (Dairy minus Vegan)
H_2O	1,125 liters (297 US gal)	383 liters (101 US gal)	374 liters (99 US gal)	113 liters (30 US gal)	1,021 liters (270 US gal)	**171 liters of water saved per serving (45 US gal)**
M	8.3 kg (18.3 lbs)	1.4 kg (3.1 lbs)	0	0	9.7 kg (21.3 lbs)	**1.6 kg (3.5 lbs)**
GHG	3.01 kg CO_{2e}	0.71 kg CO_{2e}	1.3 kg CO_{2e}	.375 kg CO_{2e}	2.1 kg CO_{2e}	**.35 kg CO_{2e} (Driving 1.3 km or 0.8 miles)**

H_20 = Water Footprint, M =Manure produced, GHG = Greenhouse Gas

Nutritional Benefits per Serving

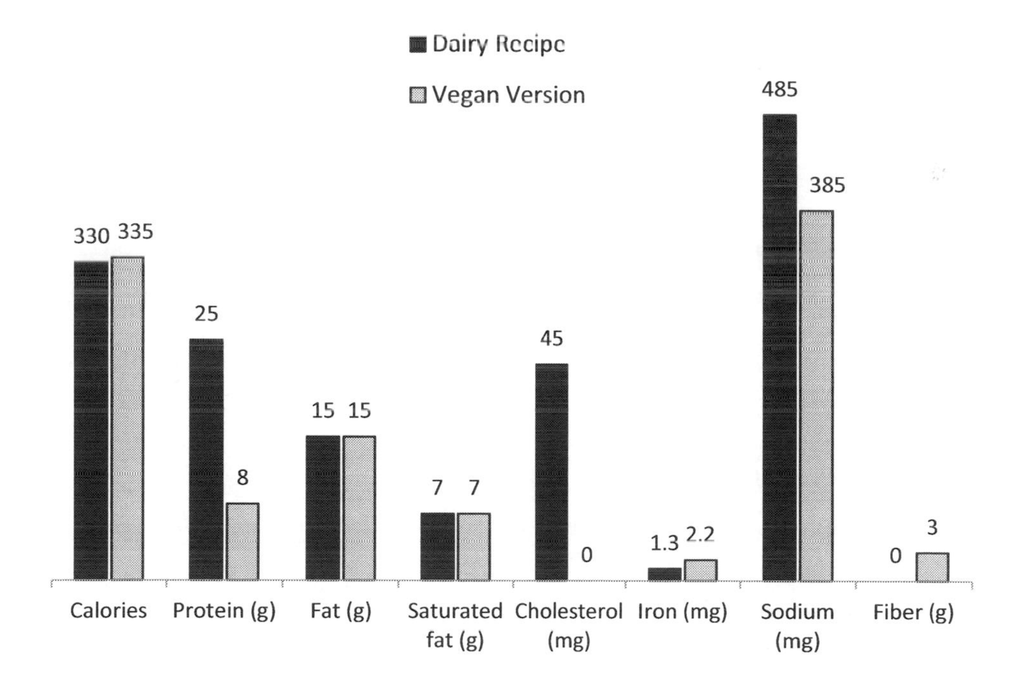

Blueberry Cinnamon Muffins (12)

Traditional Recipe	Vegan Version
2 cups all-purpose flour	2 cups (500 ml) whole-wheat or gluten-free flour
1 Tbsp baking powder	1 Tbsp (15 ml) baking powder
½ cup sugar	½ cup (125 ml) sugar
2 tsp cinnamon	2 tsp (10 ml) cinnamon
¾ cup milk	*1 cup (250 l) plant-based milk**
½ cup shortening	½ cup (125 ml) canola oil
2 eggs	—
1 cup blueberries	1 cup (250 ml) blueberries
1 tsp vanilla extract	1 tsp (5 ml) vanilla extract

**Soy milk was used for the calculations. You may wish to substitute rice, oat, or coconut milk, each of which offers a different flavor and texture.*

Instructions:

1. Preheat oven to 350 °F (180 °C).
2. Combine the first 4 (dry) ingredients together in a large bowl.
3. In a separate bowl, mix together the remaining ingredients.
4. Pour the wet mixture into the dry mixture, and stir until just blended.
5. Spoon into greased muffin pan and bake for 20 minutes or until done.

In Jonathan Balcombe's 2010 article "The Inner Lives of Animals," published in Psychology Today, he reported a Cambridge University experiment in which heifers (under a year old) experienced "Eureka" moments when they discovered how to press a panel with their noses to open a gate for food. Similarly, the Daily Mail reported that a heifer belonging to a dairy farmer in Northern Ireland deduced how to open a metal gate with her tongue in order to reach the pasture.

Environmental Savings

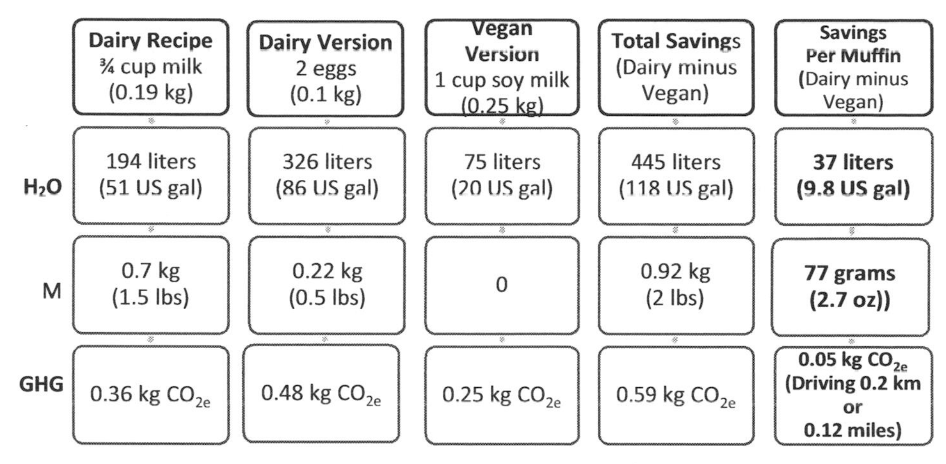

	Dairy Recipe ¾ cup milk (0.19 kg)	Dairy Version 2 eggs (0.1 kg)	Vegan Version 1 cup soy milk (0.25 kg)	Total Savings (Dairy minus Vegan)	Savings Per Muffin (Dairy minus Vegan)
H_2O	194 liters (51 US gal)	326 liters (86 US gal)	75 liters (20 US gal)	445 liters (118 US gal)	**37 liters (9.8 US gal)**
M	0.7 kg (1.5 lbs)	0.22 kg (0.5 lbs)	0	0.92 kg (2 lbs)	**77 grams (2.7 oz))**
GHG	0.36 kg CO_{2e}	0.48 kg CO_{2e}	0.25 kg CO_{2e}	0.59 kg CO_{2e}	**0.05 kg CO_{2e} (Driving 0.2 km or 0.12 miles)**

H_2O = Water Footprint, M =Manure produced, GHG = Greenhouse Gas

Nutritional Benefits per Serving

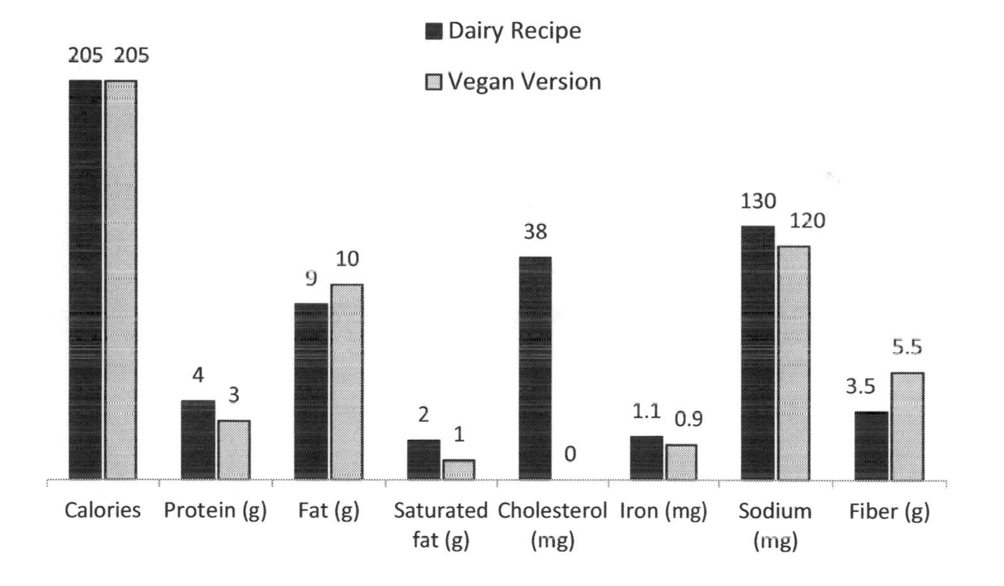

Banana Bread (12 slices)

Traditional Recipe	Vegan Version
2 cups all-purpose flour	2 cups (500 ml) whole-wheat flour
¾ cup brown sugar	¾ cup (185 ml) brown sugar
1 tsp cinnamon	1 tsp (5 ml) cinnamon
1 Tbsp baking powder	1 Tbsp (15 ml) baking powder
1 tsp baking soda	1 tsp (5 ml) baking soda
½ cup oil	½ cup (125 ml) oil
½ cup milk	*1 cup (250 ml) plant-based milk**
2 eggs	—
1 tsp vanilla extract	1 tsp (5 ml) vanilla extract
1 tsp lemon juice	1 Tbsp (15 ml) lemon juice
4 ripe bananas, mashed	4 ripe bananas, mashed
1 cup walnuts (optional)	1 cup (250 ml) walnuts (optional)

**Soy milk was used in the calculations. You may wish to substitute rice, oat, or coconut milk, each of which offers a different flavor and texture.*

Instructions:

1. Preheat oven to 350 °F (180 °C).
2. Combine the flour, sugar, cinnamon, baking powder, and baking soda.
3. In a large, separate bowl, combine the plant-based milk, oil, vanilla extract, lemon juice, and mashed bananas. Whisk until well mixed.
4. Add the dry mixture to the wet ingredients, until just blended. Fold in walnuts (if using). Pour into a baking dish and bake for about 50 minutes, until an inserted toothpick comes out clean.

FACT: Dairy cows are deprived of the opportunity to nurse their young, who are taken away 1 to 2 days after birth. Male offspring are confined to crates for 4 to 5 months, after which time they are killed for their anaemic flesh (known as veal). These babies never know their mothers, who themselves are impregnated once a year, over a period of 4 to 5 years until they are deemed no longer profitable. They are then slaughtered at an age that represents one-fifth of their natural life span of 20 years.

Environmental Savings

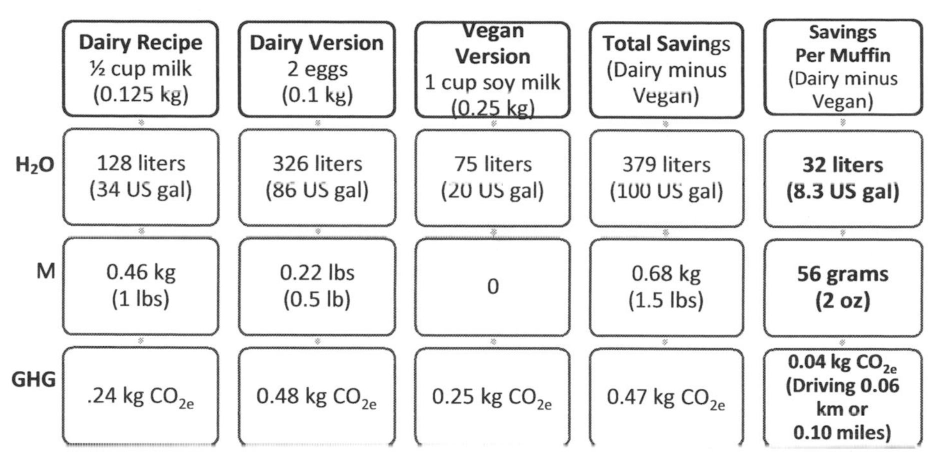

	Dairy Recipe ½ cup milk (0.125 kg)	Dairy Version 2 eggs (0.1 kg)	Vegan Version 1 cup soy milk (0.25 kg)	Total Savings (Dairy minus Vegan)	Savings Per Muffin (Dairy minus Vegan)
H_2O	128 liters (34 US gal)	326 liters (86 US gal)	75 liters (20 US gal)	379 liters (100 US gal)	**32 liters (8.3 US gal)**
M	0.46 kg (1 lbs)	0.22 lbs (0.5 lb)	0	0.68 kg (1.5 lbs)	**56 grams (2 oz)**
GHG	.24 kg CO_{2e}	0.48 kg CO_{2e}	0.25 kg CO_{2e}	0.47 kg CO_{2e}	**0.04 kg CO_{2e} (Driving 0.06 km or 0.10 miles)**

H_2O = Water Footprint, M =Manure, GHG = Greenhouse Gas

Nutritional Benefits per Serving

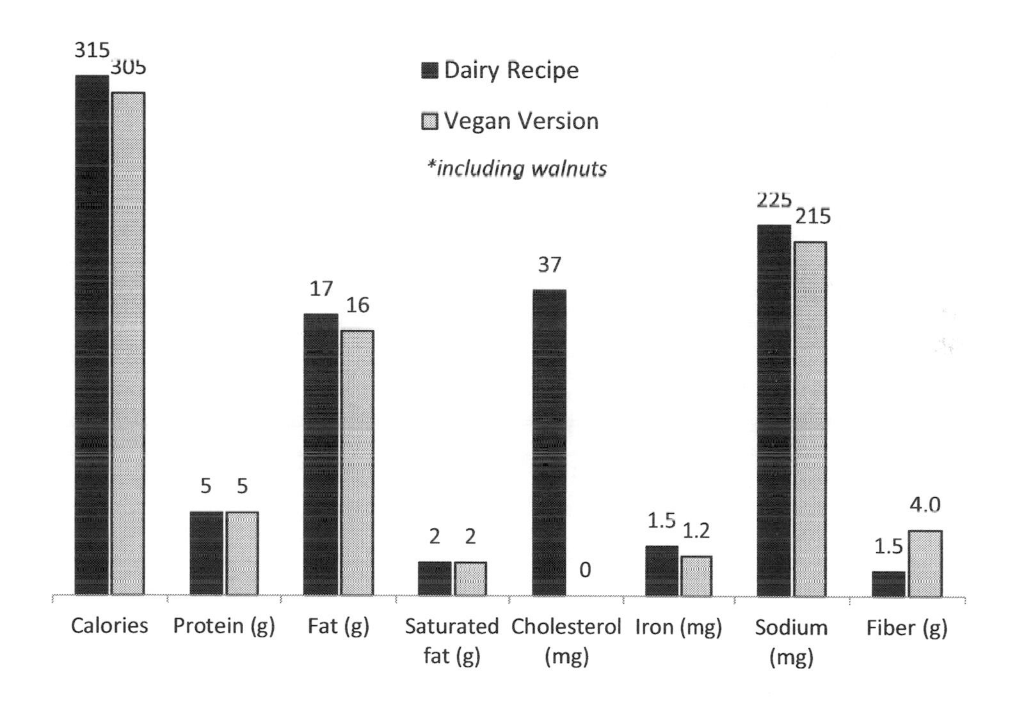

SUMMARY

Environmental Savings

(Dairy minus vegan)

Environmental Parameter (per serving)	Simple Cheese Scones	Mac and Cheese	Blueberry Cinnamon Muffins	Banana Bread
Water saved	128 liters (34 US gal)	171 liters (45 US gal)	37 liters (9.8 US gal)	32 liters (8.3 US gal)
Manure saved	1.15 kg (2.5 lbs)	1.6 kg (3.5 lbs)	77 grams (2.7 oz)	56 grams (2 oz)
GHG saved, in kg CO_{2e}	0.26 ⇔ driving 1 km (0.6 miles)	0.35 ⇔ driving 1.3 km (0.8 miles)	0.05 ⇔ driving 0.2 km (0.12 miles)	0.04 ⇔ driving 0.06 km (0.1 mile)

CONCLUSIONS

Environmental Savings

i. Significant environmental savings are achieved in vegan recipes, particularly in those using use cheese substitutes. As the recipes indicate, the water footprint savings alone range from an impressive 128 liters (34 US gallons) per cheese scone to 171 liters (45 US gallons) per serving of Mac and Cheese.

ii. Manure savings range from 1.15 kg (2.5 lbs) per cheese scone to 1.6 kg (3.5 lbs) per serving of Mac and Cheese.

iii. GHG savings range between 0.26 kg of CO_{2e} for one cheese scone to 0.34 kg for one Mac and Cheese serving. These savings are equivalent, in the first instance, to driving 1 km (0.6 mile); and, in the second instance, to driving 1.3 km (0.8 mile).

iv. The vegan versions of recipes that did not originally include cheese resulted in lesser but still notable environmental savings: 32 liters (8.3 US gallons) of water per slice of banana bread, and 37 liters (9.8 US gallons) of water per muffin.

v. Manure and GHG savings for these latter recipes are less significant than the water savings yielded. However, at roughly 110 grams (4 oz) of manure produced for each Grade A large egg (Appendix, Table 2), a 3-egg omelette would result in 330 grams (12 oz) of manure.

SUMMARY

Nutritional Benefits

(Dairy minus vegan)

Nutritional Parameter (per serving)	Simple Cheese Scones	Mac and Cheese	Blueberry Cinnamon Muffins	Banana Bread
Calories	20 fewer	5 more	0 (identical)	10 fewer
Protein	6 g less	17 g less	1 g less	0 (identical)
Fat	4 g less	0 (identical)	1 g more	1 g less
Saturated fat	4 g less	0 (identical)	1 g less	0 (identical)
Cholesterol	70 mg eliminated	45 mg eliminated	38 mg eliminated	37 mg eliminated
Iron	0.3 mg more	0.9 mg more	0.2 mg less	0.3 mg less
Sodium	80 mg more	100 mg less	10 mg less	10 mg less
Fiber	3.5 g more	3 g more	2 g more	2.5 g more

CONCLUSIONS

Nutritional Benefits

i. As described at the beginning of this chapter, dairy products have a detrimental impact upon our health. The vegan Cheese Scones recipe reduced the saturated fat of the original version by almost half, and added 3.5 g of fiber where none existed. The vegan Mac and Cheese boasts 3 g of fiber more than its traditional counterpart; the Muffins and Banana Bread each provided a minimum of 2 g more fiber. Moreover, all recipes completely eliminated the cholesterol. Thus, favorite dishes may be enjoyed with slight adaptations that result in nutritional benefits, and in even greater environmental savings.

ii. Non-dairy milks are not only for vegans; they are perfect for anyone who is lactose-intolerant or allergic to dairy. Choose products with limited sugar and without GMO ingredients. A number of vegan cheese products (such as Daiya cream cheeses) are much like conventional dairy products in taste and texture.

SECTION II

The Other 99 Percent

"It's not a requirement to eat animals; we just choose to do it. So it becomes a moral choice, and one that is having a huge impact on the planet: using up resources and destroying the biosphere."

- James Cameron, movie director, environmentalist

Chapter 7
What Are Humans Designed to Eat?

"This is a message to all those out there who think that you need animal products to be fit and strong. Almost two years after becoming vegan, I am stronger than ever before and I am still improving day by day. Don't listen to those self-proclaimed nutrition gurus and the supplement industry trying to tell you that you need meat, eggs and dairy to get enough protein. There are plenty of plant-based protein sources and your body is going to thank you for stopping feeding it with dead food. Go vegan and feel the power!"

– Patrick Baboumian, strongman

Is meat a necessary or important part of a human diet? A comparison of our anatomical structures and physiological features to those of carnivores, omnivores, and herbivores will prove illuminating.

Table 7.1a

	Carnivore	Omnivore	Herbivore	Human
Facial Muscles	Reduced, to permit a wide-opening jaw	Reduced	Complex	Complex
Jaw Joint	On the same plane as molars	On the same plane as molars	Above the plane of molars	Above the plane of molars
Jaw Motion	Shearing; minimal side-to-side	Shearing; minimal side-to-side	No shearing; good side-to-side and front-to-back	No shearing; good side-to-side and front-to-back
Jaw Muscles	Temporalis	Temporalis	Masseter and Pterygoids	Masseter and Pterygoids
Mouth to Head Ratio	Large	Large	Small	Small
Teeth (incisors)	Short, pointed	Short, pointed	Broad, flat, spade-shaped	Broad, flat, spade-shaped
Teeth (canines)	Long, sharp, curved	Long, sharp, curved	Dull, short or long (for defense), or none	Short, blunt
Teeth (molars)	Sharp, jagged, blade-shaped	Sharp blades and /or flat	Flat with cusps	Flat with nodular cusps
Chewing Capacity	None (swallows food whole)	Swallows food whole, and/or by simple crushing	Extensive chewing required	Extensive chewing required

Source: "The Comparative Anatomy of Eating," Milton Mills, MD

Carnivores (such as lions) and omnivores (such as bears) have facial muscles that enable them to open their jaws wide in order to clamp down on and to tear apart their prey. Conversely, herbivores (such as giraffes) have more complex facial muscles and a small mouth opening compared to their head size. We, too, have well-developed facial muscles, as evinced by our many facial expressions, and a small mouth opening. Just compare our mouths to those of dogs and cats.

The jaw joints of carnivores and omnivores are arranged like hinges on the same plane as their molar teeth, to allow for strong clamping motion. In contrast, the jaw joints of herbivores sit abovc the plane of teeth, thus permitting greater mobility for chewing, as is the case with humans.

Carnivores and omnivores have good shearing jaw motion but minimal side-to-side motion, while herbivores have no shearing motion at all, but good side-to-side and front-to-back motion. Cats (carnivores) and dogs (omnivores) swallow their food whole, without chewing, whereas herbivores such as rabbits and deer chew extensively before swallowing. Of course, we also chew our food.

Cats (carnivores) and dogs (omnivores) swallow their food whole, without chewing, whereas herbivores such as rabbits and deer chew extensively before swallowing. Of course, we also chew our food.

Our short, blunt canine teeth bear no resemblance to those of carnivores.

The incisors of carnivores and omnivores are short and pointed, designed for grasping and shredding flesh; those of herbivores are broad, flat, and spade-shaped. Human incisors are likewise broad, flat, and spade-shaped.

Canine teeth on carnivores and omnivores are long, sharp, and curved for stabbing and tearing prey. In herbivores, these teeth are blunt and can be small, absent, or prominent for defensive purposes only. Our short, blunt canine teeth bear no resemblance to those of carnivores.

Molars in carnivores and omnivores have jagged edges, whereas those in herbivores are flat, with cusps. Human molars are rather square and also have nodular cusps.

Table 7.1b

	Carnivore	Omnivore	Herbivore	Human
Saliva	No digestive enzymes	No digestive enzymes	Carbohydrate digestive enzymes	Carbohydrate digestive enzymes
Type of Stomach	Simple	Simple	Simple, or multiple chambers	Simple
Stomach Acidity	pH ≤ 1 with food in stomach	pH ≤ 1 with food in stomach	4 ≤ pH ≤ 5, with food in stomach	4 ≤ pH ≤ 5, with food in stomach
Stomach Capacity	60 to 70% of digestive tract's volume	60 to 70% of digestive tract's volume	≤ 30% of digestive tract's volume	21 to 27% of digestive tract's volume
Small Intestine	3 to 6 times body length	4 to 6 times body length	10 to 12+ times body length	10 to 11 times body length
Colon	Simple, short, smooth	Simple, short, smooth	Long, complex, may be sacculated	Long, sacculated
Liver	Can detoxify Vitamin A	Can detoxify Vitamin A	Cannot detoxify Vitamin A	Cannot detoxify Vitamin A
Kidney	Extremely concentrated urine	Extremely concentrated urine	Moderately concentrated urine	Moderately concentrated urine
Nails	Sharp claws	Sharp claws	Flat nails, or blunt hooves	Flat nails

Source: "The Comparative Anatomy of Eating," Milton Mills, MD

Neither carnivores nor omnivores have digestive enzymes in their saliva, while herbivores and humans both have carbohydrate digestive enzymes. Stomach acidity in carnivores and omnivores has a pH of 1 or less, strong enough to kill pathogens residing in raw flesh. Yet the stomach acidity of herbivores has a pH between 4 and 5—and again, the same is true of humans.

Because carnivores and omnivores swallow their prey in chunks, their stomachs comprise 65 percent of their digestive tract's volume. In contrast, the herbivores' pre-digestive chewing of their food requires a stomach size that is less than 30 percent of the digestive tract's volume. Human stomach capacity is even lower, at 25 percent; it is not designed for storing animal body parts.

The length of the small intestines in carnivores and omnivores is only three to six times their body length; in herbivores, it's 10 to 12 times their body length, to allow for the adequate absorption of nutrients. Similarly, our intestinal length is 10 to 11 times our body length.

Stomach acidity in carnivores and omnivores has a pH of 1 or less, strong enough to kill pathogens residing in raw flesh. Yet the stomach acidity of herbivores has a pH between 4 and 5—and again, the same is true of humans.

Carnivores and omnivores have simple, short, and smooth colons. The colons of herbivores are long, complex, and may be sacculated (divided into numerous pouches). The human colon also is long and sacculated.

Carnivores and omnivores can detoxify vitamin A in their livers, while herbivores cannot. Nor can humans.

Finally, carnivores (such as cats) and omnivores (such as raccoons) have sharp claws, while herbivores (such as cows and elephants) have flattened nails or blunt hooves. Humans have flattened nails.

Overall, the striking number of features shared by humans and herbivores–as opposed to those of carnivores or omnivores–reveals that our bodies are designed to eat plants, not animal flesh.

A CONSIDERATION OF THE PALEOLITHIC DIET

The Paleolithic (commonly called the Paleo) Diet is essentially a low-carb, high-protein diet centered on meat, fish, eggs, greens, fruits, and nuts. Anthropologist Nathaniel Dominy, Associate Professor of Anthropology and Biological Sciences at Dartmouth College, studies the relationship between the evolution of man and nonhuman primates and their diets. Addressing the view that the hunter-gatherers' diet was meat-based, he states flatly, "That's a myth. Hunter-gatherers, the majority of their calories come from plant foods . . . meat is just too unpredictable." Craig Stanford of the University of Southern California, and Co-director of the Jane Goodall Research Center, observes, "Chimpanzees are largely fruit eaters, and [the consumption of] meat [accounts for] only about 3 percent of the time they spent eating overall." That three percent includes termites, their main non-plant food source.

PROTEIN REQUIREMENTS

The World Health Organization recommends that humans obtain only 5.5 to 13 percent of their calories from protein, although certain conditions such as advanced age, bodybuilding, or the healing of major wounds may require more protein. The Centers for Disease Control and Prevention recommend 56 and 46 grams of protein respectively for men and women over the age of 19. This is based upon the RDA daily recommendation of 0.8 g of protein per kg of body weight. Thus, a man who weighs 70 kg would require about 56 grams of protein. Yet the National Health and Nutrition Board survey of 2013 found that the average American male consumes 102 grams of protein per day, while the average female consumes 70 grams; clearly more than is recommended or required.

Research in early 2014 from the University of Southern California signals the dangers of high-protein diets. After tracking adults for almost 20 years, researchers found that middle-aged people (50 – 65 years old) who ate an animal-rich diet were four times more likely to die of cancer than were those eating a low-protein diet. Dr. Valter Longo, Professor of Gerontology/Alzheimer's Research/Cancer Research, Biological Sciences and senior author, declared, "We provide convincing evidence that a high-protein diet—particularly if the proteins are derived from animals—is nearly as bad [for your health] as smoking . . ." The study concluded that the risk of developing cancer is almost as high as that of smoking 20 cigarettes a day. A further revealing observation is that the increased risk of death from cancer, diabetes, or other causes was either non-existent or reduced if the high-protein diet was plant-based. High-protein is defined as over 20 percent of total calories. A 300-gram (11-oz) steak alone has 77 grams of protein! The researchers pointed out that dairy products are also high in protein; for example, a 200-ml (0.8-cup) glass of milk represents 12 percent of the Recommended Dietary Allowance (RDA), and a 40-gram (1.5-oz) slice of cheese contains 20 percent of the RDA for protein. A single chicken breast or salmon fillet fulfills 40 percent of RDA protein requirements. Consequently, it is all too easy to eat too much protein when following a diet centered on meat and dairy. However, Dr. Longo also found that, for people over 65, a higher protein intake was actually associated with reduced risks of dying, a result of our body's decreased ability to absorb or to process proteins as we age.

The Centers for Disease Control and Prevention recommend 56 and 46 grams of protein respectively for men and women over the age of 19.

Yet the National Health and Nutrition Board survey of 2013 found that the average American male consumes 102 grams of protein per day, while the average female consumes 70 grams; clearly more than is recommended or required.

Dr. Valter Longo, Professor of Gerontology/Alzheimer's Research/Cancer Research, Biological Sciences and senior author, declared, "We provide convincing evidence that a high-protein diet—particularly if the proteins are derived from animals—is nearly as bad [for your health] as smoking . . ."

Anthropological evidence and a scientific comparison of our anatomical and physiological features overwhelmingly indicate that the human species is best suited to be herbivores. Therefore, it is entirely plausible that many of today's chronic diseases can be halted or reversed by means of a plant-based diet that Homo sapiens were designed to eat. Not only is a plant-based diet nutritionally adequate, but elite athletes around the world are embracing it to achieve superior performance.

Anthropological evidence and a scientific comparison of our anatomical and physiological features overwhelmingly indicate that the human species is best suited to be herbivores.

Chapter 8
Vegetables

Cream of Celery Soup
Corn and Potato Chowder
Stuffed Mushroom Caps
Vegetable Goulash
Ratatouille
Cauliflower Casserole
Classic Stir-Fry
Indian-Spiced Stir-Fry
Broccoli and Snow Pea Stir-Fry
Seven Wonders Power Salad
Nine Lives Salad
Thai Vegetable Salad
Cauliflower "Couscous" Salad
Spicy Roasted Yams
Roasted Balsamic Eggplant
Pan-Seared Brussels Sprouts in Wine
Marsala Peas and Carrots

"A global shift towards a vegan diet is vital to save the world from hunger, fuel poverty and the worst impacts of climate change."

- United Nations Environment Programme's (UNEP) International Panel of Sustainable Resource Management

Vegetables' rainbow of colors offers a visual and culinary feast. The following phytonutrient-packed recipes present various preparation methods for both common and less common vegetables. They are inspired by ethnic flavors—and are mostly gluten-free.

Cream of Celery Soup (4 – 6 servings)

Creamed vegetable soups are both a great way to warm up on a cold day and a creative way to use leftover veggies. Celery is high in fiber and a good source of magnesium, folate, vitamin K, and potassium. Soy milk adds calcium to this comforting soup.

4 cups (1 L) celery, chopped
4 small red potatoes, cubed
1 white onion, chopped
1 cup (250 ml) chopped mushrooms
2 garlic cloves, minced
1 Tbsp (15 ml) canola oil
2 Tbsp (30 ml) whole-wheat, or gluten-free flour
1 tsp (5 ml) black pepper
2 cups (500 ml) vegetable broth
2 cups (500 ml) soy milk

Instructions:

1. Bring 2 cups of broth to a boil in a soup pot, and add the chopped celery and potatoes. Reduce heat to simmer.

2. In a skillet, sauté the onions, mushrooms, and garlic in oil until soft.

3. Add the skillet contents to the soup pot. Slowly add the flour, stirring constantly. Then add in the black pepper. Simmer for 20 minutes until potatoes are done. Turn off heat.

4. Mash the soup's ingredients until crushed, then return to simmer.

5. Add the soy milk and simmer for another 10 minutes.

Corn and Potato Chowder (6 servings)

This classic favorite tastes as good as traditional versions but avoids the heart-clogging saturated fat and cholesterol of dairy products.

6 nugget potatoes, peeled and diced into small cubes
1 Tbsp (15 ml) extra-virgin olive oil, or canola oil
1 onion, coarsely chopped
4 cloves garlic, chopped
2 stalks celery, cut into small slices
2 medium carrots, cut into slices
1 tsp (5 ml) salt
1 tsp (5 ml) pepper
2 cups (500 ml) vegetable broth
2 cups (500 ml) soy milk
1½ cups (375 ml) corn kernels
2 Tbsp (30 ml) flour
1 Tbsp (15 ml) chopped parsley
Vegan bacon bits (optional)

Instructions:

1. Wash the nugget potatoes and cut them into 1-inch (2.5-cm) cubes. Set aside.

2. In a soup pot, sauté the onion and garlic in oil, on medium heat. Add the celery, carrots, salt, and pepper.

3. Add the vegetable broth, soy milk, and corn kernels. Bring to a boil. Reduce heat and simmer for 30 minutes.

4. Mix the flour with enough water to form a paste. Stir the flour paste and parsley into the soup.

5. Continue simmering for another 10 – 15 minutes, stirring occasionally, until soup has thickened slightly.

6. Sprinkle vegan bacon bits on top when serving.

Stuffed Mushroom Caps (4 – 6 servings)

Mushrooms are naturally low in sodium and calories, and contain beneficial phytonutrients as well as B-complex vitamins, selenium (an antioxidant) and potassium. With so many varieties of mushrooms on the market today, experimenting with different kinds allows for a variety of textures and flavors.

- 20 large button mushrooms, or 6 large Portobello mushrooms
- ¼ cup (60 ml) quinoa
- ½ cup (125 ml) vegetable broth
- 1Tbsp (15 ml) canola oil
- ½ cup (125 ml) minced onion
- 3 garlic cloves, minced
- 1 Tbsp (15 ml) dried parsley, or 3 Tbsp, (45ml) fresh
- ½ tsp (3 ml) oregano
- ½ tsp (3 ml) thyme
- 1 Tbsp (15 ml) nutritional yeast
- ½ cup (125 ml) Daiya shreds (optional)

Instructions:

1. Preheat oven to 400 °F (200 °C).

2. Rinse mushrooms and snap off the stems. Chop up the stems.

3. Cook the quinoa in vegetable broth by bringing it to a boil, then simmering for 10 minutes. Turn off the heat.

4. In a pan, sauté the onion, garlic, and chopped mushroom stems on medium heat for about 5 minutes.

5. Remove from heat and add the cooked quinoa, followed by parsley, oregano, thyme, and nutritional yeast.

6. Scoop the filling into the mushrooms and place them on a baking sheet.

7. If desired, top with Daiya shredded cheese.

8. Place in oven for 12 – 15 minutes.

Vegetable Goulash (4 servings)

This is another favorite household dish, especially on a cold night. The thick Portobello mushrooms and sweet carrots and parsnips impart a hearty flavor reminiscent of beef stew, without creating negative environmental and health impacts. More importantly, this dish provides significant anti-cancer and cholesterol-lowering benefits.

1 large onion, chopped
4 cloves garlic, minced
4 Portobello mushrooms, chopped
1 Tbsp (15 ml) canola oil
4 stalks celery, diced
2 carrots, chopped
2 parsnips, diced

1 red bell pepper, diced
1 Tbsp (15 ml) paprika
1 Tbsp (15 ml) caraway seeds
½ tsp (3 ml) salt
½ tsp (3 ml) black pepper
1½ cups (375 ml) marinara sauce
½ cup (125 ml) water

Instructions:

1. Heat the oil in a pot and sauté the onions, garlic, and mushrooms, about 5 minutes.

2. Add the celery, carrots, parsnips, and red bell pepper. Cook for another 5 minutes.

3. Add spices, marinara sauce, and water, and bring to boil.

4. Then reduce the heat to low and simmer for 30 minutes.

Ratatouille (4 servings)

In Europe and elsewhere, eggplants are known as "aubergines," and come in several varieties. Their purple skin is a source of anthocyanins, antioxidants that also are believed to have anti-inflammatory and heart-protective qualities. Because they offer bulk with minimal calories and virtually no fat, eggplants are excellent for weight loss regimens.

1 large onion, chopped
4 cloves garlic, minced
2 cups (500 ml) mushrooms, quartered
1 Tbsp (15 ml) canola oil
1 large eggplant, cut into cubes
1 large zucchini, thickly sliced, then halved
½ tsp (3 cm) black pepper
1 tsp (5 ml) thyme
1 tsp (5 ml) oregano
2 cups (500 ml) pasta sauce (basil, mushroom, or marinara)
½ cup (125 ml) water

Instructions:

1. Heat oil in a large pot and sauté onion, garlic, and mushrooms for about 5 minutes.

2. Add the eggplant, zucchini, and spices, and sauté over medium heat for another 10 minutes.

3. Add pasta sauce and water. Once the ratatouille mixture starts to bubble, decrease the heat and simmer for another 15 minutes.

4. Stir occasionally. If too thick, add water. If too thin, add more pasta sauce.

Cauliflower Casserole (4 servings)

This is a nutritious version of a comfort food.

1 medium cauliflower, cut into small florets
1 Tbsp (15 ml) vegan margarine
3 stalks green onion, sliced thinly
2 cloves garlic, grated
1 cup (250 ml) plant-based milk
½ tsp (3 ml) salt
½ tsp (3 ml) pepper
½ cup (125 ml) Daiya Cheddar shreds
½ cup (125 ml) Daiya Pepperjack shreds

Instructions:

1. Preheat oven to 350 °F (180 °C). Oil a casserole dish.

2. Steam cauliflower florets over a pot of boiling water until crisp-tender (still somewhat firm), about 8 minutes. Drain and set aside.

3. In a sauce pan, melt the margarine and cook the green onion and garlic for about 2 minutes.

4. Add in the plant-based milk, salt, pepper, and both kinds of Daiya shreds. Stir to combine. Cook until cheese has melted.

5. Place cauliflower florets into the prepared casserole dish, and pour the cheese mixture on top.

6. Bake for 15 minutes, covered. Then bake another 5 – 10 minutes uncovered, until slightly browned.

Classic Stir Fry (4 servings)

Stir-frying is the easiest way to use left-over veggies. If desired, organic tofu may be added to a number of these recipes to enhance their protein and calcium content. Fragrant jasmine rice is a particularly good accompaniment to this dish.

1 Tbsp (15 ml) ginger, minced
4 cloves garlic, minced
2 stalks green onion, diced
1 Tbsp (15 ml) canola oil
1 broccoli crown, cut into large florets
3 baby bok choy, or Shanghai bok choy, coarsely chopped
20 sugar snap peas, or snow peas
1 stalk celery, sliced diagonally, about ¼-inch (6-mm) thick
10 mushrooms, halved
1 medium carrot, thinly sliced
2 oz (60 grams) tofu, cut into cubes (optional)
Soy sauce, or teriyaki sauce (or gluten-free tamari sauce, if preferred)
1 tsp (5 ml) sesame oil

Instructions:

1. Heat the oil in a wok or stir-fry pan. Add ginger, garlic, and green onion. Cook for 5 minutes.

2. Lower heat to medium and add the vegetables—including cubed tofu—if desired. Cook for about 5 minutes.

3. Add teriyaki, soy, or tamari sauce, to taste. Turn off heat.

4. Drizzle with sesame oil, and stir through before serving.

Indian-Spiced Stir-Fry (4 servings)

Stir-frying (although an Asian cooking method) is used here to create an Indian-inspired dish pungent with cumin, onion, and garlic. This dish is best served with basmati rice.

1 small onion, sliced
3 cloves garlic, chopped
1½ Tbsp (22 ml) oil
2 celery ribs, diced
1 large or 2 small orange peppers, sliced into 2-inch (5-cm) stripes
2 cups (500 ml) green beans, cut into 2-inch (5-cm) sections
1 Tbsp (15 ml) cumin
1 Tbsp (15 ml) coriander
12 grape tomatoes
12 oz (340 grams) firm organic tofu, cut into cubes (optional)
3 Tbsp (45 ml) chutney, any variety
¼ tsp (1 – 2 ml) chili flakes
½ tsp (3 ml) salt (optional)
Black pepper to taste

Instructions:

1. Heat the oil in a large pot. Sauté the onion and garlic until soft.

2. Add the celery, peppers, green beans, cumin, and coriander. Sauté over medium heat for another 3 minutes.

3. Add the tomatoes, tofu (if using), chili flakes, salt, and pepper.

4. Heat through, stirring from time to time, for approximately 2 minutes.

5. Drain off liquid, and add chutney.

6. Coat thoroughly and turn off heat.

Broccoli and Snow Pea Stir-Fry (4 servings)

Stir-fry broccoli and crisp snow peas offer a satisfying crunch. Tossing in organic tofu (optional) adds extra calcium and protein.

- 1½ Tbsp (22 ml) peanut or canola oil
- 2 broccoli crowns, cut into large florets
- 2 Tbsp (30 ml) ginger, minced
- 2 stalks green onion, sliced diagonally
- 1 whole garlic bulb, minced
- 20 pieces of snow peas
- 12 oz (340 grams) firm organic tofu, cut into cubes (optional)
- 1 cup (250 ml) mushrooms, halved
- 1 tsp (5 ml) sesame oil
- ½ cup (125 ml) Hoisin sauce, or black bean sauce
- 5 sprigs cilantro for garnish
- 1 tsp (5 ml) chili pepper flakes (optional)

Instructions:

1. Heat the oil in a wok or large pot. Add the broccoli, ginger, green onion, and garlic. Stir-fry on medium heat for 3 minutes.

2. Add the snow peas, tofu (if desired), and mushrooms. Stir-fry for another 3 minutes. Vegetables should still be crunchy.

3. Reduce heat to low and drain off excess liquid.

4. Add the sesame oil and hoisin or black bean sauce. Coat the vegetables and turn off the heat.

5. Transfer to a serving dish and garnish with cilantro and chili pepper flakes (if desired).

6. Serve immediately, with brown basmati or fragrant jasmine rice.

Seven Wonders Power Salad (2 – 3 servings)

In addition to its excellent phytonutrients, this salad offers visual appeal, abundant fiber, and a satisfying crunch. A new twist is afforded here by the addition of the refreshing, slightly citrus-flavored jicama. This root vegetable, commonly referred to as "Mexican yam" or "yam bean," is native to Mexico, Central and South America, and Southeast Asia. Jicama is high in fiber, low in fat, and serves as a crisp, juicy accent when added raw. Jicama may also be stir-fried with other vegetables. It is low in calories, containing about 80 percent water, and is high in vitamin C.

1 red bell pepper, diced
1 orange bell pepper, diced
3 large radishes, thinly sliced
1 cup (250 ml) grape tomatoes, halved
1 small jicama, shredded
3 stalks celery, diced
1 large red cabbage leave, shredded; or 2 Napa cabbage leaves, thinly sliced
6 oz (170 grams) organic firm tofu, diced
Few slices of purple onion, diced finely
1 tsp (5 ml) ginger, minced
¼ cup (60 ml) ginger sesame dressing
Cilantro for garnish
Toasted sesame seeds (optional)

Instructions:

1. Combine all ingredients in a large salad serving bowl.
2. Pour dressing over mixture. Blend well.
3. Refrigerate for 30 minutes before serving.
4. Garnish with cilantro and toasted sesame seeds.

Nine Lives Salad (4 servings)

This delightfully colorful salad is another favorite in our household. Calcium and iron-rich kale is complemented by peppers rich in vitamin C, accompanied by the crunch of walnuts and radishes, and topped with zucchini and tantalizing black olives.

1 large yellow or orange bell pepper, diced
7 medium radishes, thinly sliced (or 2 ribs celery, if preferred)
1 cup (250 ml) grape tomatoes, halved
5 kale leaves, de-ribbed and finely chopped
½ cup (125 ml) fresh or frozen (thawed) cranberries, halved
½ cup (125 ml) walnut pieces, quartered, or pecan halves
1 bunch fresh basil leaves, chopped
1 garlic scape (tender garlic shoot, if seasonally available), or garlic clove
I small zucchini, diced
½ cup (125 ml) sliced black olives
5 Tbsp (75 ml) Italian-type dressing

Instructions:

1. Combine all but the last three ingredients in a large salad bowl.
2. Pour dressing over mixture. Blend well.
3. Top with diced zucchini and sliced olives.
4. Chill for 30 minutes before serving.

Thai Vegetable Salad (4 servings)

This marvelously nutritious salad, with peanuts and bean sprouts, has an oriental twist. Beans or peas may be added on top for extra protein.

½ bunch broccoli (about ¾ lb, or 340 grams)
1 cup (250 ml) bean sprouts
½ red medium bell pepper, chopped
3 green onions (white and green tender parts), sliced
1 Tbsp (15 ml) sesame oil
5 Tbsp (75 ml) peanuts, coarsely crushed
¼ cup (60 ml) seasoned rice vinegar
2 Tbsp (30 ml) low-sodium, non-GMO soy sauce
1 Tbsp (15 ml) dry sherry
¼ tsp (1 – 2 ml) cayenne
2 tsp (10 ml) freshly ground ginger

Instructions:

1. Pulse the broccoli in a food processor for a few seconds.

2. Pour into large salad bowl. Add the bean sprout, red pepper, and green onion and combine.

3. To make the dressing, combine all the remaining ingredients in a small bowl. Be sure to use seasoned rice vinegar as opposed to unseasoned.

4. Pour the dressing over the vegetables, and marinate for 1 hour.

Cauliflower "Couscous" Salad (4 – 6 servings)

Rich in vitamins and antioxidants, cauliflower is used in this recipe to mimic the texture and appearance of couscous. The cauliflower florets are transformed into the size of couscous pearls in seconds, with a few pulses in the food processor.

1 medium cauliflower, cut into florets
1 yellow bell pepper, coarsely cut
2 garlic cloves, minced
2 green onions, sliced
1 Tbsp (15 ml) grated fresh ginger
2 Tbsp (30 ml) apple cider vinegar
2 tsp (10 ml) curry powder
1 tsp (5 ml) cinnamon
1 Granny Smith apple, diced
¼ cup (60 ml) slivered almonds
¼ cup (60 ml) raisins or cranberries
½ cup (125 ml) chopped fresh mint
5 Tbsp (75 ml) low-fat Thai dressing

Instructions:

1. Place the cauliflower florets in a food processor, and pulse until you have obtained the size of couscous. Repeat with bell pepper.

2. In a large bowl, combine the garlic, green onion, ginger, vinegar, curry powder, and cinnamon.

3. Add the cauliflower couscous mixture, followed by the apple, almonds, raisins or cranberries, and mint. Mix well.

4. Pour dressing over the "couscous" salad, and toss to coat.

Spicy Roasted Yams (4 servings)

Orange-fleshed yams (as distinct from the white-fleshed variety) are a great source of beta-carotene. When cooked, they are as soft and sweet as canned pumpkin. They are not to be confused with sweet potatoes, which are dense and starchy when cooked. Yams are readily available in Canada. In parts of the US, sweet potatoes are sometimes labeled as "yams."

2 large orange-fleshed yams
½ cup (125 ml) balsamic vinegar
½ cup (125 ml) olive oil
1 tsp (5 ml) thyme (dried)
1 tsp (5 ml) oregano (dried)
2 tsp (10 ml) parsley (dried)
½ tsp (3 ml) cayenne pepper
½ tsp (3 ml) black pepper
2 Tbsp dried chives (30 ml) (optional)
Cold water

Instructions:

1. Wash the yams. Leaving the skins on, cut them into quarters (or smaller sections, if using large yams).

2. Place them in a large pot, and add water to cover. Bring to a boil.

3. Reduce heat to simmer and cook for 20 – 30 minutes, until almost tender. Drain and set aside. Remove skins.

4. Preheat oven to 350 °F (180 °C).

5. Mix together the remaining ingredients. Add enough water to make 1¼ cup (310 ml) liquid. Set aside.

6. Place the cooked yams in a baking dish.

7. Pour the balsamic mixture over the yams, and bake in the oven for 20 – 30 minutes, until liquid is absorbed.

Roasted Balsamic Eggplant (6 servings)

Simply marinate and bake these eggplants to savor their earthy taste and to benefit from the phytonutrients packed in their purple skins. Eating this vegetable improves blood circulation and brain and heart health, while its fiber content helps in the management of diabetes.

4 medium eggplants, peeled and cut into ½-inch (1.3-cm) cubes
2 tsp (10 ml) salt
6 medium shallots, or 1 large red onion
2 garlic cloves, minced
½ cup (125 ml) balsamic vinegar
¼ cup (60 ml) extra-virgin olive oil
½ tsp (3 ml) black pepper
1 bunch fresh basil, finely chopped

Instruction:

1. Preheat oven to 400 °F (200 °C).
2. Place eggplant cubes in a large colander and sprinkle with salt, and drain for 30 minutes, tossing once or twice.
3. Dry the eggplant with paper towels.
4. Prepare the marinade by combining onion, garlic, vinegar, oil, and black pepper.
5. Toss the eggplant to coat well.
6. Place on a baking sheet and roast until soft, about 30 minutes. Toss with a spoon every 15 minutes, to ensure even cooking
7. Remove from the oven, let cool, and toss with basil.

Pan-Seared Brussels Sprouts in Wine (6 servings)

Even people who shy away from Brussels sprouts are enticed by the aroma and tangy flavor of this dish. Brussels sprouts have the highest glucosinolate concentration of all the cruciferous vegetables, yielding significant anti-cancer protective qualities. They also offer anti-inflammatory support.

2 Tbsp (30 ml) canola oil
3 cloves garlic, minced
4 cups (1 L) Brussels sprouts, ends trimmed off, split in half
½ cup (125 ml) white wine
3 Tbsp (45 ml) chopped pecan
Salt and pepper to taste

Instructions:

1. Heat the oil in a skillet and pan-fry the garlic and brussel sprouts.

2. Reduce to low heat and add wine as needed to prevent skillet from drying out.

3. Continue cooking until sprouts are tender but still firm.

4. Season with salt and pepper. Top the dish with chopped pecans.

Marsala Peas and Carrots (4 servings)

This recipe imbues peas and carrots with an aromatic spice and a boost of lycopene (an antioxidant) from tomatoes. While frozen peas and carrots may be used, fresh carrots will provide better texture.

1 medium purple onion, chopped
3 garlic cloves, chopped
1 Tbsp (15 ml) canola oil
2 stalks celery, diced
3 large carrots, sliced ½-inch (1.3-cm) thick
1 Tbsp (15 ml) Marsala spice
1 tsp (5 ml) curry powder
1 tsp (5 ml) black pepper
1 tsp (5 ml) cayenne, or chili flakes (optional)
3 cups (750 ml) peas, thawed
1½ cups (375 ml) marinara sauce

Instructions:

1. Heat the oil in a large pot, and add onion and garlic. Cook over medium heat, until soft.

2. Add the celery and carrots. Add a bit of water, to prevent sticking; then add the spices, and cook for another 10 minutes.

3. Add the peas and pasta sauce. Turn up heat until the mixture starts to bubble.

4. Reduce heat to low and simmer for 5 minutes, stirring occasionally.

Chapter 9
Pulses and Grains

Split Pea and Potato Soup

Curried Lentil and Pasta Soup

Cabbage and Navy Bean Soup

Simple Bean and Vegetable Stew

Yam and Black Bean Chili

Quick Barbecued Beans

Veggie Bean Burgers

Mexican Breakfast

Cajun Bean and Rice

Barley Pilaf

Asian Noodle Salad

Quinoa with Fennel and Walnut Salad

Quinoa and Bean Casserole

Sprouted-Grain Mini Pizzas

One day the absurdity of the almost universal human belief in the slavery of other animals will be palpable. We shall then have discovered our souls and become worthier of sharing this planet with them.

–Martin Luther King, Jr.

According to Statistics Canada, almost one third of children are either overweight or obese. Furthermore, 60 percent of Canadian men and 45 percent of Canadian women are either overweight or obese. Similarly, the US Centers for Disease Control report that close to 70 percent of American adults are either overweight or obese, while over one third of American children are in the same predicament. These are sobering numbers. The importance of incorporating pulses in our daily diets, therefore, cannot be overemphasized.

Dried beans, split peas, lentils, and chickpeas are collectively known as pulses. There is a wonderful array of bean varieties available. While kidney beans, black beans, pinto beans, and chickpeas are the best known, the flavors and nutritional benefits of navy beans, lima beans, fava beans, Great Northern beans, broad beans, aduki beans, mung beans, cannellini beans, and black-eyed peas should not be overlooked. As well, lentils come in black, green, and red varieties.

Pulses are protein powerhouses. These nutritional staples are high in iron, potassium, calcium, folate, zinc, and magnesium, as well as soluble and insoluble fiber that enhance digestive health. Pulses are also extremely low in fat. Regular consumption of beans, lentils, split peas, and chickpeas contributes significantly to the control of chronic diseases and to weight management by lowering cholesterol, triglycerides, and blood pressure.

Additionally, pulses play a key role in the treatment of metabolic syndrome. One third of American adults are affected by this cluster of conditions, the characteristic symptoms being glucose intolerance, abdominal obesity, hypertension, and dyslipidemia (a disorder of lipoprotein metabolism).

If you are not accustomed to eating pulses, introduce them into your diet, a ½-cup (125-ml) serving of lentils or split peas every few days. As you gradually increase this frequency, begin adding beans. Any bloating or flatulence generally will subside within the first two weeks. Once your body has adapted, it is recommended that beans, lentils, or split peas be included on a daily basis.

You may choose to purchase dried beans (which require soaking but are more economical), or to buy canned beans for convenience. To reduce their cooking time and to aid digestion, soak dried beans for eight to ten hours, then rinse until all the foam is washed away, before cooking. Smaller beans cook more quickly than larger beans. While the sodium content of many brands of canned beans is high, low-sodium brands (containing fewer than 150 mg per serving) are generally available. Even canned beans should be rinsed so as to remove approximately one third of the sodium and much of the compound responsible for digestive discomfort.

Pulses are extremely versatile, and are featured in many ethnic cuisines. They may be added to salads or stews, mashed with vegetables and baked into burger patties, or blended into hummus (ground chick-pea spread). They may be eaten hot or cold, as a garnish, or as a main dish. To your health!

Split Pea and Potato Soup (4 servings)

This hearty soup, packed with fiber and rich in antioxidants, is splendid on a wintery day. Add a piece of whole grain bread and a green salad to create a complete meal.

2 cups (500 ml) yellow or green split peas, washed (no pre-soaking required)
1 Tbsp (15 ml) canola oil
1 small onion, chopped
4 cloves garlic, minced
6 cups (1.5 L) vegetable broth
1 medium potato, peeled and cubed
Salt and pepper to taste

Instructions:

1. Rinse the split peas well. Drain and set them aside.
2. Heat the oil in a large soup pot and cook the onion and garlic until the onion is soft.
3. Pour the vegetable broth into the soup pot and add the split peas, potatoes, salt, and pepper.
4. Bring the soup to a boil, then simmer for 2 hours, or until split peas are cooked.
5. Stir well to break up any lumps before serving.

Curried Lentil and Pasta Soup (4 servings)

If you're looking for a satisfying meal that doesn't require much preparation time, this hearty soup will fit the bill.

- 1 Tbsp (15 ml) oil
- 1 cup (250 ml) chopped purple onion
- 3 celery stalks, diced
- 3 cloves garlic, minced
- 1 large carrot, diced
- 1 medium zucchini, diced
- 4 cups (1 L) vegetable broth
- 1 can (6 oz, or 170 grams) tomato paste
- 2 cups (500 ml) cooked lentils
- 4 oz (110 grams) whole wheat macaroni pasta (for gluten-free version, substitute brown-rice or buckwheat pasta)
- 4 kale pieces, de-ribbed, and coarsely chopped
- 1 Tbsp (15 ml) curry powder

Instructions:

1. Heat the oil in a large pot. Add the onions, celery, garlic, carrot, and zucchini. Cook for about 5 minutes.

2. Add the vegetable broth and tomato paste. Stir until paste is well mixed into broth, and bring to a boil.

3. Reduce the heat and add the lentils.

4. Add the pasta, kale, and curry powder.

5. Simmer for another 15 minutes, or until pasta is cooked.

Cabbage and Navy Bean Soup (6 servings)

Beans and cabbage are a great combination in soups. This recipe lends itself well to the use of different beans and different cabbages (red, green, or savoy) each time.

1 medium cabbage, chopped
1 Tbsp (15 ml) canola oil
1 medium onion, chopped
4 cloves garlic, chopped
14-oz (400-gram) can stewed tomatoes
5 cups (1.25 L) vegetable broth
2 tsp (10 ml) oregano
3 cups (750 ml) cooked navy beans
Salt and pepper

Instructions:

1. Core the cabbage and coarsely chop into large pieces.
2. Heat the oil in a large soup pot and sauté the onion and garlic for 5 minutes.
3. Place the chopped cabbage in the large soup pot. Add the stewed tomatoes and broth.
4. Add the oregano, beans, salt, and pepper.
5. Bring to a boil, then simmer for 30 minutes.

Simple Bean with Vegetable Stew (4 servings)

This one-pot meal uses collard greens in addition to traditional stewing vegetables. Collards, related to kale and cabbage, are high in vitamins A, C, K, as well as some B-complex vitamins. They also are good sources of iron, calcium, and manganese.

- 1 Tbsp (15 ml) canola oil
- 1 small onion, chopped
- 3 cloves garlic, diced
- ½ cup (125 ml) vegetable broth
- 3 collard leaves, de-stemmed, diced
- 2 cups (500 ml) mixed vegetables
- 28-oz (800-ml) can of beans, rinsed and drained
- 1 can (14 oz, or 400 ml) seasoned, diced tomatoes
- 1 Tbsp (15 ml) Italian blend seasoning
- Salt and pepper to taste

Instructions:

1. In a large pot, sauté the onion and garlic in oil, until onion is soft.
2. Add broth, collards, and mixed vegetables. Bring to a boil, then reduce heat to simmer.
3. Add the beans, diced tomatoes, and seasoning. Simmer for 20 minutes.

Yam and Black Bean Chili (6 – 8 servings)

The contrasting colors of orange and black give this unexpected taste combination great visual appeal.

2 cups (500 ml) yams, peeled and chopped into several big pieces
1 Tbsp (15 ml) canola oil
1 medium onion, chopped
4 cloves garlic, minced
1 large bell pepper, chopped
[3] cans of 14-oz (400-ml)) black beans, rinsed and drained
1 Tbsp (15 ml) chili powder
1 cup (250 ml) vegetable broth
15-oz (420-gram) can crushed tomatoes
Cayenne, to taste

Instructions:

1. In a small pot, cook the yam pieces until tender. Drain and set aside.
2. In a large soup pot, sauté the onions, garlic, and bell peppers in oil for about 5 minutes.
3. Add all the remaining ingredients except cayenne to the soup pot, including the cooked yams.
4. Simmer for about 20 minutes, or until the chili thickens. Add cayenne for extra pizazz.
5. Serve over brown rice.

Quick Barbecue Beans (6 servings)

This is a great replacement for canned baked beans, which usually contain excessive amounts of sodium and sugar. This recipe reduces the usual sugar content by 30 percent and the sodium by more than half. The result is a flavorful, nutrient-dense dish that costs about half as much as its commercial counterpart.

1 Tbsp (15 ml) canola oil
2 small onions, chopped
1 whole garlic bulb, minced
1 cup (250 ml) tomato paste
2 cups (500 ml) water
4 Tbsp (60 ml) ketchup
1 tsp (5 ml) cinnamon
1Tbsp (15 ml) soy or tamari sauce
1 tsp (5 ml) hot chili flakes (optional)
4 cups (1 L) cooked pinto beans
½ tsp (3 ml) cayenne or ground chipotle pepper
2 Tbsp (30 ml) balsamic vinegar

Instructions:

1. In a large stock pot, heat the oil, and sauté the onion and garlic until soft.
2. Add all the remaining ingredients. Combine until well mixed.
3. Bring to a boil, then simmer for 20 minutes.

Veggie Bean Burgers (4 burgers)

These mushroom-and-bean burgers hold up well when pan-grilled. Slide them onto your favorite burger bun, top with lettuce, tomato, relish, onion, mushrooms, and mustard to dress it up.

1 small red or orange bell pepper
1 small onion
1 cup (250 ml) mushroom
1 can (14-oz or 400-ml) black or kidney beans, drained
¼ cup (60 ml) chili sauce
½ tsp (3 ml) salt
½ tsp (3 ml) pepper
½ tsp (3 ml) garlic powder
½ cup (125 ml) rolled oats or wheat germ (or gluten-free flour)
Oil for pan-frying

Instructions:

1. Place the vegetables in a food processor, and pulse until finely chopped.
2. Add the beans, chili sauce, salt, pepper, garlic powder, and rolled oats. Pulse until well mixed.
3. Chill the mixture for 1 hour, then form into patties.
4. Pan-fry in oil for about 5 minutes, until patties are golden brown with crisp edges.

Mexican Breakfast (3 – 4 servings)

An alternative to the popular grand slam breakfast, this version is made with protein-packed beans, accompanied by diced potatoes. Offering fiber and zing (from the chili powder) and no cholesterol, this will keep you going happily until lunch.

1 Tbsp (15 ml) canola oil
1 medium onion, diced
1 medium red bell pepper, diced
1 cup (250 ml) mushrooms, chopped
½ tsp (3 ml) salt
2 cups (500 ml) cooked cannelloni beans
2 cups (500 ml) diced potatoes, cooked
½ tsp (3 ml) garlic powder
½ tsp (3 ml) paprika
½ tsp (3 ml) chili powder
Salsa (for garnish)

Instructions:

1. Have ready 2 cups of cooked diced potatoes. You can buy frozen ones or cook a large potato in a pot of water. Drain, cool, then dice.

2. In a large skillet, sauté the onions, bell pepper, mushrooms, and salt in oil, until vegetables are soft and all the liquid has evaporated.

3. Add the cooked potatoes and the remaining ingredients except salsa, and continue cooking until heated through.

4. Serve with salsa.

Cajun Bean and Rice (8 servings)

The combination of beans and rice is traditional in many cultures. This spicy version of red beans has the added kick of cayenne pepper.

4 cups (1 L) cooked brown rice
1 Tbsp (15 ml) canola oil
1 cup (250 ml) chopped mushrooms
1 large yellow or orange bell pepper, diced
½ cup (125 ml) vegetable broth
4 cups (1 L) cooked kidney beans, rinsed and drained
1 Tbsp (15 ml) Cajun seasoning
½ tsp (3 ml) salt
½ tsp (3 ml) ground cayenne pepper
1 tsp (5 ml) garlic powder

Instructions:

1. Cook the brown rice according to package instructions. Set aside.
2. In a large pot, heat the oil and sauté the mushrooms and bell pepper until liquid has evaporated. Add remaining ingredients.
3. Simmer for 20 minutes.
4. Serve over cooked brown rice.

Barley Pilaf (4 servings)

An ancient grain with high fiber content, barley improves blood-sugar control and digestive regularity, and also lowers cholesterol. Additionally, it is a good source of niacin and selenium.

1 Tbsp (15 ml) canola oil
1 small onion, chopped
2 cloves garlic, minced
1 red bell pepper, chopped
3 cups (750 ml) vegetable broth
1 cup (250 ml) barley
1 cup (250 ml) mixed vegetables
1 tsp (5 ml) black pepper
Fresh basil or cilantro, chopped

Instructions:

1. In a large pot, sauté the onion in olive and garlic in oil until soft.
2. Add in the remaining ingredients except for basil or cilantro.
3. Bring to a boil, uncovered. Then reduce heat, cover, and simmer for about 35 – 40 minutes.
4. Garnish with fresh basil or cilantro as desired.

Asian Noodle Salad (4 servings)

This noodle salad is ideal on a hot summer day, and can be made ahead for picnics or dinner on the deck. The crunchy celery and radish slices and refreshing tomatoes nicely balance the chewy spaghettini.

- 6 medium radishes, thinly sliced
- 3 stalks celery, thinly sliced
- 1 medium red, orange, or yellow bell pepper, diced
- 2 stalks green onion, or ¼ of a small purple onion, sliced thinly
- 10 grape tomatoes, halved
- ½ package (½ lb, or 225 grams) whole-wheat spaghettini, or any gluten-free noodle
- 1 can (14-oz, or 400-ml) black beans, rinsed and drained
- 3 Tbsp (45 ml) rice vinegar
- 1 Tbsp (15 ml) sesame oil
- 4 Tbsp (60 ml) light ginger dressing

Instructions:

1. Bring 4 cups of water to a boil.
2. Meanwhile, chop the vegetables.
3. Break the raw spaghettini straws into halves before adding them to the boiling water. Cook according to package directions.
4. Drain the spaghettini and transfer it to a large mixing bowl.
5. Add the black beans and chopped vegetables to the bowl, and mix well.
6. Add rice vinegar and sesame oil, and mix again.
7. Add ginger dressing, and mix again.
8. Chill for 1 hour before serving.

Quinoa with Fennel and Walnut Salad (4 servings)

High-protein quinoa is very easy to make because it cooks in 10 minutes. Fennel is high in vitamin C and potassium, while walnuts are high in omega-3 fatty acid, making this salad a winning combination of antioxidants.

1 cup (250 ml) quinoa
2 cups (500 ml) vegetable broth
1 small fennel, chopped
10 grape tomatoes, halved
1 small yellow bell pepper, chopped
¼ cup (60 ml) diced purple onion
1 tsp (5 ml) thyme
½ tsp (3 ml) sage
Pepper to taste
¼ cup (60 ml) walnut pieces
¼ cup (60 ml) Italian dressing
¼ cup (60 ml) fresh parsley, chopped

Instructions:

1. Cook the quinoa in vegetable broth by bringing it to a boil. Immediately reduce the heat and simmer for 10 – 12 minutes. Remove from heat and let cool.

2. Chop the fennel, tomatoes, yellow bell pepper, and onion.

3. Mix together in a medium bowl with the thyme, sage, pepper, and walnut pieces.

4. Add the cooled quinoa to the vegetable mixture and blend thoroughly.

5. Coat with dressing. Garnish with parsley. Chill for 30 minutes.

Quinoa and Black Bean Casserole (6 servings)

This quickly prepared, colorful dish may be modified to include whatever vegetables are available. It is a complete meal, with protein, vegetables, and quinoa, and can be enhanced with Daiya cheese.

1 cup (250 ml) quinoa
2 cups (500 ml) water
1 can (14-oz, or 400-ml) black beans
1 zucchini, diced
1 orange bell pepper, diced
3 Tbsp (45 ml) green olives (optional)
1 cup (250 ml) grape tomatoes, halved
1 tsp (5 ml) chili powder
1 tsp (5 ml) oregano
Salt and pepper to taste
1 Tbsp (15 ml) oil
1 cup (250 ml) mild salsa
1 cup (250 ml) shredded Daiya cheese (optional)

Instructions:

1. Preheat oven to 350 °F (180 °C). Cook the quinoa in 2 cups (500 ml) of water. Bring to a boil and simmer for 12 minutes. Remove from heat.

2. Drain and rinse the black beans.

3. In a large bowl, combine the quinoa and black beans.

4. Add the vegetables and spices. Mix together. Add oil and mix again.

5. Pour half the salsa onto the bottom of a casserole dish. Transfer the vegetables and the quinoa-bean mixture from the bowl into the casserole dish.

6. Pour the remaining salsa on top. Bake for 30 minutes.

7. If desired, top with shredded Daiya vegan cheese and heat for another 12 minutes, or until melted.

Sprouted-Grain Mini Pizzas (4 slices)

You can make the healthiest pizza imaginable in less time than it takes to have one delivered to your door. Sprouted-grain breads have a higher protein content than regular whole-grain breads. They also have a denser texture, making them great for use as a pizza crust. Daiya cheese is a healthier, cheaper, and environmentally friendlier option than conventional cheeses in commercial pizzas.

4 slices spouted-grain bread
2 vegan sausages (optional)
¾ cup (185 ml) pasta sauce
1 tsp (5 ml) garlic powder
1 tsp (5 ml) dried oregano
1 cup (250 ml) mixed field greens
½ cup (125 ml) black olives
½ - 1 cup (125 – 250 ml) Daiya shredded vegan cheese

Instructions:

1. Toast the sprouted grain bread lightly in a toaster.
2. Preheat the oven to 350 °F (180°C). If using vegan sausages, slice them thinly.
3. Spread a layer of pasta sauce and sausages (if desired) onto each toasted bread slice.
4. Sprinkle garlic powder and oregano, then add greens and olives.
5. Top with vegan shredded cheese.
6. Place the mini pizzas in a preheated oven on wire racks for 15 minutes, until the vegan cheese has melted. Ensure that air is circulating beneath the toasts, to produce the desired crispness.

Chapter 10
Soy (Tofu)

Spicy Baked Tofu

Potato and Tofu Curry

Tofu and Bean Nachos

Classic Tofu Stir-Fry

Lettuce Wraps

Calzones

Vegan Sausage with Broccoli

Potato Salad with Vegan Sausages

Tofu Stew with Root Vegetables

Crusted Tempeh with Orange-Mustard Sauce

One-Pot Rice with Tofu

Vegan Lasagne

Until we extend our circle of compassion to all living things, humanity will not find peace."

~Albert Schweitzer, The Philosophy of Civilization

Spicy Baked Tofu (4 servings)

This is a great introductory tofu dish for those who are not familiar with this food. This will become a regular staple once you've tasted it. It can be serve hot with rice or cold as an appetizer.

1 lb (450 grams) extra-firm organic tofu
1 Tbsp (15 ml) minced ginger
1 Tbsp (15 ml) minced garlic
¼ cup (60 ml) soy sauce
2 tsp (10 ml) Sriracha chili sauce, or other chili sauce
2 Tbsp (30 ml) maple or agave syrup
2 Tbsp (30 ml) ketchup
1 Tbsp (15 ml) rice vinegar
2 Tbsp (30 ml) sesame seeds (optional)
Black pepper to taste
1 tsp (5 ml) oil for coating casserole dish

Instructions:

1. Preheat oven to 375 °F (190 °C). Lightly oil a baking casserole.

2. Cut the tofu block in half, then press out excess water gently. Cut into 1-inch (2.5-cm) cubes.

3. In a medium bowl, combine the minced ginger, garlic, soy sauce, Sriracha chili sauce, syrup, ketchup, and rice vinegar.

4. Coat the tofu pieces in the marinade, and let stand for 10 minutes.

5. Place the tofu cubes in a single layer in the prepared casserole dish. Sprinkle sesame seeds on top. Add black pepper as desired.

6. Bake in the oven for about 20 – 25 minutes.

Potato and Tofu Curry (4 servings)

This is a twist on a traditional curry dish. The tofu will soak up the flavor of the curry sauce. Basmati rice would be an excellent accompaniment.

4 medium potatoes
1 lb (450 grams) extra-firm organic tofu, cubed
1 Tbsp (15 ml) canola oil
1 Tbsp (15 ml) minced ginger
1 Tbsp (15 ml) minced garlic
1 red onion, sliced
1½ cup (375 ml) diced tomatoes
1 Tbsp (15 ml) curry powder
2 tsp (10 ml) cumin
½ tsp (3 ml) salt
1 tsp (5 ml) black pepper
Cilantro, for garnish

Instructions:

1. Cook the potatoes in a pot of water until done, but still firm.
2. Let cool, then peel and cut into 1-inch (2.5-cm) cubes. Set aside.
3. Cut the tofu block in half horizontally, then press out excess water gently. Cut into 1-inch (2.5-cm) cubes.
4. In a medium pot, sauté in oil, the ginger, garlic, and onion for about 5 minutes, until onion is soft.
5. Add the diced tomatoes, curry powder, cumin, salt, and pepper. Mix well in the pot. Bring to a gentle boil and reduce heat to simmer.
6. Add the potato and tofu cubes; mix well into the curry. Simmer for 15 minutes.
7. Garnish with cilantro just before serving.

Tofu and Bean Nachos (3 – 4 servings)

The addition of tofu to this bean nacho increases the appetizer's protein content.

½ lb (225 grams) firm organic tofu
1 Tbsp (15 ml) canola oil
1 onion, chopped
3 stalks green onion, chopped
2 garlic cloves, minced
1 tsp (5 ml) cumin
½ tsp (3 ml) oregano
1½ cups (375 ml) pineapple salsa
14-oz can (400-ml) black beans, rinsed and drained
2 green chilies, diced
13-oz (360-gram) bag restaurant-style, non-GMO tortilla chips
2 cups (500 ml) Daiya shredded Pepper Jack cheese
4 cups (1 L) shredded romaine lettuce
1 cup (250 ml) guacamole
2 Tbsp (30 ml) cilantro, chopped

Instructions:

1. Cut the tofu into ½-inch (1.3-cm) cubes.
2. Preheat the oven to 375 °F (190 °C).
3. In a skillet, cook the onion, green onions, garlic, and tofu cubes in oil for about 5 minutes.
4. Reduce heat and add the cumin, oregano, and salsa.
5. Add black beans and chilies. Simmer until sauce thickens, about 10 minutes.
6. Place tortilla chips in a baking dish. Sprinkle Daiya shreds on top. Bake until "cheese" melts, about 5 minutes.
7. Transfer tortilla chips from the baking dish to a serving platter. Surround with shredded lettuce.
8. Spoon the skillet mixture over the melted Daiya cheese. Spoon guacamole over the skillet mixture.
9. Top with cilantro.

Classic Tofu Stir-Fry (4 servings)

Stir-fry is a great way to use whatever vegetables you have in your fridge: everything from cruciferous vegetables to leafy greens may be substituted for the suggestions below. Choosing a different sauce each time—whether black bean, hoisin, Szechwan, garlic stir-fry, or even curry sauce—will result in wonderful versatility.

1 Tbsp (15 ml) canola oil
1 Tbsp (15 ml) ginger, minced
4 cloves garlic, minced
2 stalks green onion, sliced
1 package (12 – 13 oz) firm or extra-firm organic tofu (about 350 grams), cut into cubes
1 broccoli crown, cut into large florets
3 baby bok choy, or Shanghai bok choy, coarsely chopped
20 snap peas
1 stalk celery, sliced diagonally, about ¼-inch (6-mm) thick
6 – 8 mushrooms, halved
2 – 3 medium carrots, thinly sliced
soy sauce or teriyaki sauce (or gluten-free tamari sauce)
1 tsp (5 ml) sesame oil for finishing

Instructions:

1. Heat the canola oil in a wok or stir-fry pan on medium.
2. Add the ginger, garlic, and green onion. Cook for 5 minutes.
3. Add the tofu and vegetables. Cook for about 5 minutes.
4. Add soy, teriyaki, or tamari sauce, to taste. Turn off the heat.
5. Stir, and drizzle with sesame oil.

Lettuce Wraps (4 servings)

Most lettuce wraps are filled with meat and are high in fat. This delicious version is much more healthful, and far better for the environment.

11 – 12 oz (340 grams) extra-firm organic tofu, crumbled
1 Tbsp (15 ml) oil
3 garlic cloves, minced
1 Tbsp (15 ml) ginger, minced
5 medium mushrooms, sliced
1 stalk celery, thinly sliced
1 carrot, finely diced
½ tsp (3 ml) ground black pepper
3 Tbsp (45 ml) Hoisin sauce
1 head of lettuce, leaves separated and washed
2 stalks green onion, thinly sliced
½ cup (125 ml) chopped cilantro

Instructions:

1. Coarsely crumble the tofu with a masher.
2. In a large skillet, heat the oil and cook the tofu, garlic, ginger, mushrooms, celery, and carrot for 5 – 7 minutes. Drain off excess liquid.
3. Add pepper and Hoisin sauce. Mix well and remove from heat.
4. Garnish a platter with green onion and cilantro, accompanied by a plate of washed lettuce leaves with which to make the wraps.
5. Transfer the hot filling onto the garnished platter and enjoy.

Calzones (4 servings)

This is a more healthful and eco-friendly version of the traditional, meat- and dairy-based calzones.

1 lb (450 grams) firm organic tofu, crumbled
1 Tbsp (15 ml) canola oil
1 onion, diced
4 garlic cloves, minced
1 cup (250 ml) kale, torn
1 small bell pepper, diced
1 cup (250 ml) pasta sauce
½ cup (125 ml) red wine
½ tsp (3 ml) black pepper
½ tsp (3 ml) oregano
¼ tsp (1 – 2 ml) chili pepper flakes (optional)
1 cup (4 oz), or 250 ml (110 grams) Daiya Mozzarella shreds
14-oz (400-gram) refrigerated pizza crusts
½ cup (125 ml) plant-based milk

Instructions:

1. Pulse the firm tofu in a food processor, or mash with a potato masher, into pea-size pieces.

2. Preheat the oven to 350 °F (180 °C).

3. In a skillet, heat the oil and cook the onion, garlic, kale, bell pepper, and tofu crumble for about 5 minutes. Drain off excess liquid.

4. Add the pasta sauce, wine, and spices. Cook for another 7 minutes, until sauce thickens. Turn off the heat.

5. Add Daiya Mozzarella shreds and stir. Remove the filling from the heat.

6. Unroll the pizza crusts and cut them into 4 squares.

7. Spoon ¼ of the cooked filling into the center of each square. Brush the edges with soy milk.

8. Fold each square diagonally, and press the edges together to seal.

9. Bake for 15 – 20 minutes, or until crusts are golden brown.

Vegan Sausage with Broccoli (4 servings)

Substituting vegan sausages for meat-based ones will reduce your intake of saturated fat and cholesterol. It also will reduce substantial water usage, eliminate manure production and minimize greenhouse gas emissions.

1 Tbsp (15 ml) canola oil
2 garlic cloves, minced
2 broccoli crown, cut into florets
1 large bell pepper, chopped
10 mushrooms, quartered
1 carrot, sliced
4 links vegan sausages, sliced
½ cup (125 ml) pasta sauce
Pepper to taste

Instructions:

1. Sauté the garlic, broccoli florets, bell pepper, mushrooms, and carrots in oil over medium heat until lightly browned for 2 – 3 minutes.

2. Add the vegan sausages and pasta sauce.

3. Cook for another 5 minutes, or until sausages are warmed.

4. Serve over whole-wheat spaghetti or gluten-free pasta.

Potato Salad with Vegan Sausages (2 – 3 servings)

Potato salad is a summertime staple. Instead of the cholesterol-laden meat sausages that often accompany them, try a healthy vegan sausage made with organic tofu or wheat gluten.

8 nugget potatoes
2 links of vegan sausages
2 stalks celery, diced
Black or green olives, sliced (optional)
2 stalks green onion, sliced
Veganaise (vegan mayonnaise), sufficient to coat ingredients
Salt and pepper to taste

Instructions:

1. Wash the nugget potatoes.
2. Cook them in boiling water until tender, but still firm. Drain and let cool.
3. Cut them into bite-sized chunks.
4. Cut the vegan sausages into ½-inch (1.3-cm) slices, and place in a toaster oven at 350 °F (180 °C). Bake for 10 minutes. Let cool.
5. Combine all ingredients and mix them well with Veganaise.
6. Season with salt and pepper, to taste.
7. Refrigerate for at least 1 hour before serving.

Tofu Stew with Root Vegetables (4 servings)

Stews are very "forgiving" dishes to prepare. The tofu in this dish will soak up the savory flavor imparted by the vegetables and spices.

1 Tbsp (15 ml) canola oil
1 onion, chopped
2 carrots, chopped
2 celery ribs, chopped
2 purple top turnips, chopped
2 parsnips, chopped
1 lb (450 grams) extra-firm organic tofu, cut into bite-size pieces
2 cups (500 ml) vegetable broth
1 cup (250 ml) tomato paste
1 tsp (5 ml) oregano
1 tsp (5 ml) thyme
1 tsp (5 ml) marjoram
1 tsp (5 ml) black pepper
1Tbsp (15 ml) corn starch

Instructions:

1. In a pot, sauté the onion in oil until translucent.

2. Add the remaining ingredients except for the tofu and corn starch.

3. Bring to a boil and simmer at low heat for 30 minutes, until root vegetables are cooked.

4. Add the tofu pieces. Dissolve the corn starch in some of the cooking liquid and mix well into the entire pot.

5. Simmer for another 10 minutes, until sauce thickens slightly.

Crusted Tempeh with Orange-Mustard Sauce (3 servings)

Tempeh is a fermented form of soy. Many of the available products are already marinated. For instance, Tofurky brand offers Smokin' Bacon, Coconut Curry, and Sesame Garlic flavors. Flavored varieties need only to be heated in a toaster oven or added directly to a stir-fry. If you can't find marinated tempeh, try the following recipe.

8 oz (225 grams) tempeh
3 Tbsp (45 ml) low-sodium soy sauce
1 Tbsp (15 ml) water
½ - ¾ cup (125 – 375 ml) corn flake crumbs, bread crumbs, panko
1 tsp (5 ml) garlic powder
1 tsp (5 ml) onion powder
1 tsp (5 ml) paprika
¼ cup (60 ml) orange marmalade
1½ Tbsp (22 ml) yellow or spicy brown mustard
1 Tbsp (15 ml) orange juice

Instructions:

1. Slice the tempeh in half, crosswise, then into about ½-inch (1.3-cm) strips. Set aside.

2. In a small bowl, combine soy sauce and water. Set aside.

3. On a large plate, add the spices to the corn flake crumbs and mix well. Set aside.

4. In another bowl, combine the marmalade, mustard, and juice. Set aside.

5. Preheat the oven to 350 °F (180 °C).

6. Dredge the tempeh strips into soy sauce, then coat all sides with crumb mixture.

7. Arrange on a foil-lined cookie sheet and gently spoon the remaining sauce over the tempeh strips. Alternatively, it may be used as a dipping sauce.

8. Bake for 30 minutes, or until the sauce topping has thickened.

One-Pot Rice with Tofu (4 servings)

This is a flavorful, quick, and easy meal. While waiting for the rice to cook, cut up some raw vegetables to serve with a dip.

1 Tbsp (15 ml) oil
1 medium onion, chopped
2 cloves garlic, minced
1 Tbsp (15 ml) ginger, minced
1 lb (450 grams) firm organic tofu, cubed
1 cup (250 ml) long-grain jasmine or basmati rice
2 Tbsp (30 ml) soy sauce
10 small button mushrooms, halved
2 cups (500 ml) vegetable broth

Instructions:

1. In a medium pot, sauté the onion, garlic, and ginger in oil for 2 – 3 minutes.

2. Add the remaining ingredients and cook for about 30 minutes, until water is absorbed.

Vegan Lasagna (6 servings)

This lasagne uses whole-wheat noodles and has zero cholesterol. It is destined to become a family favorite.

1 lb (450 grams) firm organic tofu, crumbled
4 cups (1 L) pasta sauce
1 Tbsp (15 ml) dried basil or 3 fresh leaves, chopped
2 tsp (10 ml) oregano
6 strips whole-wheat or gluten-free lasagna noodles
2 cups (500 ml) vegan ricotta
2½ cups (625 ml) shredded Daiya "Mozzarella"
¼ cup (60 ml) hot water

Vegan ricotta (2 cups)

15 ounces (420 grams) firm organic tofu
½ tsp (3 ml) garlic powder
½ tsp (3 ml) onion powder
1½ Tbsp (22 ml) fresh lemon juice
2 tsp (10 ml) nutritional yeast flakes

Instructions for ricotta:

1. Put the tofu in a blender, and blend until lumpy.
2. Add remaining ingredients and blend again. Set aside.

Cooking instructions for lasagne:

1. Crumble the tofu with a masher, to pea-size crumble. Sauté the crumbled tofu in a skillet with pasta sauce and spices.

2. Preheat the oven to 350 °F (180 °C). Make the vegan ricotta.

3. Cover the bottom of an 11-inch (28-cm) by 7-inch (18-cm) baking dish with one third of the pasta-tofu sauce.

4. Place 3 lasagna noodles on top to form the next layer. Follow with half of the vegan ricotta and half of the Daiya Mozzarella shreds.

5. Repeat: Pour on another layer of pasta-tofu sauce (one third), followed by 3 noodles, then by the remaining vegan ricotta and Daiya shreds.

6. Spread the last third of the pasta-tofu sauce on top. Pour hot water around the edges. Cover with aluminum foil and bake for 45 minutes. Uncover and bake for another 15 minutes.

Chapter 11
Baking Without Dairy or Eggs

Cheese and Corn Muffins

Jam Muffins

Simple Blueberry Scones

Cinnamon Pancakes

Vegetable Frittatas

Peanut Butter Cookies

Apricot Coconut Squares

Brownies

Simple Chocolate Cake

Sweet Potato and Apple Pie

Spinach and Mushroom Quiche

"The assumption that animals are without rights and the illusion that our treatment of them has no moral significance is a positively outrageous example of Western crudity and barbarity. Universal compassion is the only guarantee of morality."

~Schopenhauer

Although eggs are commonly used in baking, in fact many recipes do not require them. In other recipes, eggs may be replaced with a variety of healthful products. For a comprehensive list of egg replacements, please visit the website http://chefinyou.com/egg-substitutes-cooking/

Plant-based milk alternatives constitute one of the fastest-growing segments in the food market. As mentioned in Chapter 6, choices include soy (organic recommended), rice, hemp, oat, coconut, and nut milks, all of which are produced without generating manure. If you can make delicious baked goods without generating excessive environmental impacts, without consuming saturated fat and cholesterol, and without imposing a life of suffering and death on animals, why not? The initial inconvenience of learning where plant-based milks are located at your favorite supermarket is well worth the benefits to the environment, your health, and the animals.

Cheese & Corn Muffins (1 dozen)

Who doesn't like cheese and corn? This winning combination will especially please those who are lactose-intolerant.

1 "flax egg" [1 Tbsp (15 ml) ground flaxseed plus 3 Tbsp (45 ml) water], if desired
1 cup (250 ml) all-purpose flour
½ cup (125 ml) whole-wheat flour
½ cup (125 ml) cornmeal
2 tsp (10 ml) baking powder
½ tsp (3 ml) salt
¼ cup (60 ml) sugar
1½ cups (375 ml) Daiya cheddar shreds
½ cup (125 ml) canola oil
¾ cup (185 ml) plant-based milk

Instructions:

1. Preheat the oven to 400 °F (200 °C). Grease muffin tins.
2. If using a flax egg, prepare by mixing the ground flaxseed with water in a bowl, and let stand for 15 minutes. Otherwise, omit this step. In a different bowl, mix the dry ingredients together. Add Daiya shreds and mix again.
3. After the flax egg has gelled, add the oil and milk, then mix the liquid ingredients together. (Otherwise, just combine oil and milk in a bowl different from the dry ingredients.)
4. Pour the wet ingredients into the dry mixture, and blend thoroughly.
5. Spoon into greased muffin tins, and bake for about 20 minutes.

Jam Muffins (I dozen)

Muffins are great snacks on daytrips or hikes. These jam muffins are so moist that you won't notice the lack of eggs.

2 cups (500 ml) whole-wheat or gluten-free flour
1Tbsp (15 ml) baking powder
1 tsp (5 ml) baking soda
½ cup (125 ml) sugar
1 tsp (5 ml) cinnamon
1 cup (250 ml) plant-based milk
½ cup (125 ml) canola oil
2 tsp (10 ml) apple-cider vinegar
1 tsp (5 ml) vanilla extract
½ cup (60 ml) raspberry, strawberry, apricot, or other jam of your choice

Instructions:

1. Preheat the oven to 375 °F (190 °C). Line the muffin tins.
2. Combine the dry ingredients in a large bowl.
3. In a separate bowl, combine the liquid ingredients (all the remaining ingredients except the jam).
4. Pour the wet mixture into the dry mixture, and blend together thoroughly.
5. Fill each muffin cavity ¾ full.
6. Make an indentation in the center of each and spoon 2 tsp (10 ml) of jam into each muffin.
7. Pour the remaining batter over the tops of each muffin.
8. Bake for 20 minutes, or until slightly browned.

Simple Blueberry Scones (6 medium)

These scones are easy to make and are the perfect pairing for tea or coffee. They freeze well and, upon thawing, can be reheated in a toaster oven.

2 cups (500 ml) whole-wheat flour
½ cup (125 ml) sugar
1 tsp (5 ml) cinnamon
½ tsp (3 ml) cream of tartar
1 Tbsp (15 ml) baking powder

1 Tbsp (15 ml) lemon juice
½ cup (125 ml) oil
1 cup (250 ml) plant-based milk
1 cup (250 ml) blueberries
Icing sugar (optional)

Instructions:

1. Preheat the oven to 350 °F (180 °C).
2. In a large bowl, combine all the dry ingredients and mix well.
3. In a separate bowl, whisk together the oil, lemon juice, and plant-based milk. Add in blueberries and mix again.
4. Pour the liquid mixture into the dry ingredients, and fold together.
5. Form into 6 individual scones, and arrange them on a cookie sheet.
6. Bake for 20 – 25 minutes.
7. After the scones are cooled, sprinkle with icing sugar if desired.

Cinnamon Pancakes (4 pancakes)

Pancakes are perfect for breakfast on a lazy weekend. These cinnamon pancakes are so fluffy that you'll be surprised by the use of whole wheat flour, and without the use of eggs. Try different plant-based milks to impart different flavors to these healthful pancakes.

1 cup (250 ml) whole-wheat or gluten-free flour
1 Tbsp (15 ml) baking powder
1 Tbsp (15 ml) sugar
1 tsp (5 ml) cinnamon
1 cup (250 ml) plant-based milk
1 Tbsp (15 ml) canola oil
1 tsp (5 ml) vanilla extract
Cooking spray

Instructions:

1. Preheat a lightly oiled skillet.

2. Mix the dry ingredients together in a large bowl.

3. Combine the wet ingredients in a smaller bowl, and whisk them together.

4. Pour the wet ingredients into the dry mixture. Beat well.

5. Pour ¼ cup batter into the skillet, and cook on low heat until the batter rises and bubbles start to form.

6. Flip and cook the other side until golden brown. Repeat until batter is used up.

Vegetable Frittatas (4 servings)

I enjoy serving these frittatas to our guests; they are always enthusiastically received.

1 Tbsp (15 ml) canola oil
1 cup (250 ml) diced zucchini
1 medium red or sweet yellow onion, diced
2 – 3 cloves garlic, minced
1 large red bell pepper, diced
2 – 3 medium tomatoes, diced

12 oz (340 grams) firm organic tofu
½ cup (125 ml) plant-based milk
½ tsp (3 ml) salt
½ tsp (3 ml) black pepper
½ tsp (3 ml) turmeric
1 Tbsp (15 ml) nutritional yeast
1 cup (250 ml) Daiya Cheddar shreds

Instructions:

1. Preheat the oven to 375 °F (190 °C).

2. Sauté the zucchini, onion, garlic, bell pepper, and tomatoes in oil for about 7 minutes, until vegetables are softened.

3. Remove from heat and drain off the excess liquid.

4. In a blender or food processor, blend together the tofu, plant-based milk, salt, pepper, turmeric, and nutritional yeast.

5. Pour into a large bowl.

6. Fold the sautéed vegetables into the large bowl of processed tofu mixture.

7. Pour the entire mixture into an oiled baking dish or quiche pan, and bake for 40 minutes, or until the frittata is firm.

8. Top with Daiya shreds and bake for another 5 – 10 minutes, until "cheese" has melted.

Peanut Butter Cookies (2 dozen)

The peanut butter in this recipe can easily be replaced with other nut butters for variety, or if one has peanut allergy. This is a favorite for kids and the young at heart.

1½ cups (375 ml) whole-wheat flour
1 cup (250 ml) rolled oats
1 cup (250 ml) brown sugar
½ cup (125 ml) oil
½ cup (125 ml) plant-based milk
¾ cup (185 ml) peanut butter

Instructions:

1. Preheat the oven to 350 °F (180 °C).
2. In a large bowl, combine the flour, oats, and brown sugar.
3. Pour in the oil and plant-based milk, and mix well together.
4. Fold in the peanut butter and combine again. If the batter is too dry, add a little plant-based milk.
5. Shape the mixture into balls, and place them on a cookie sheet.
6. Flatten them with the back of a fork.
7. Bake for 20 – 25 minutes. Let cool.

Apricot Coconut Squares (12 – 16 servings)

This recipe may be varied with the substitution of other flavored jams or preserves.

1½ cups (375 ml) whole-wheat flour
1 cup (250 ml) quick-cooking rolled oats
¾ cup (375 ml) unsweetened coconut shreds
1 Tbsp (15 ml) baking powder
¾ cup (185 ml) sugar
¾ cup (185 ml) oil
½ cup (125 ml) plant-based milk
¾ cup (185 ml) apricot jam
2 Tbsp (30 ml) coconut milk

Instructions:

1. Preheat the oven to 350 °F (180 °C).

2. In a large bowl, combine the flour, oats, coconut shreds, baking powder, and sugar.

3. In a separate bowl, whisk together the oil and plant-based milk.

4. Pour the liquid mixture into the dry ingredients, and combine together until crumbly.

5. Place half of the crumbly mixture in a baking dish.

6. In a small bowl, combine the jam and coconut milk, mixing them with a fork until the consistency allows for easy spreading onto the crumbly layer.

7. Spread the diluted Jam on top of the crumbly layer in the baking dish.

8. Add the remaining crumbly mixture on top of the jam.

9. Bake for 45 minutes, or until done. Cut into squares.

Brownies (9 squares)

These brownies are lighter than most. So go ahead; indulge in a second helping!

½ cup (125 ml) oil
¼ cup (60 ml) plant-based milk
1 tsp (5 ml) vanilla extract
1 cup (250 ml) whole-wheat flour
½ cup (125 ml) cocoa
1 cup (250 ml) sugar
1 Tbsp (15 ml) baking powder
1 tsp cinnamon

Instructions:

1. Preheat the oven to 350 °F (180 °C). Oil an 8-inch (20-cm) square baking dish.

2. Whisk together the oil, plant-based milk, and vanilla extract in a large bowl.

3. In a separate bowl, combine the flour, cocoa, sugar, baking powder, and cinnamon.

4. Gradually blend the dry mixture into the wet one and fold.

5. Pour into the square baking dish, and bake for 30 minutes.

Simple Chocolate Cake (12 servings)

This versatile, vegan cake is easy to make, and perfect for coffee, lunch boxes, or dessert.

2 cups (500 ml) whole-wheat flour
1 Tbsp (15 ml) baking powder
2 tsp (10 ml) baking soda
1 cup (250 ml) sugar
¼ cup (60 ml) cocoa powder
1½ cups (375 ml) plant-based milk

½ cup (125 ml) canola oil
1 Tbsp (15 ml) lemon juice
2 tsp (10 ml) vanilla extract
1 cup (250 ml) chopped walnuts (optional)

Instructions:

1. Preheat the oven to 350 °F (180 °C). Oil an 8-inch (20-cm) square baking dish.

2. In a large bowl, combine the flour, baking powder, baking soda, sugar, and cocoa powder.

3. In a separate bowl, whisk together the plant-based milk, oil, lemon juice, and vanilla extract.

4. Add the wet ingredients into the dry mixture and fold until blended.

5. Pour the batter into the baking dish. If using walnuts, sprinkle on top. Bake for 40 minutes, until an inserted toothpick comes out clean.

Sweet Potato and Apple Pie (6 servings)

This is a healthy alternative to the traditional, sugar-coated sweet potato dishes commonly served at Thanksgiving. Sweet potatoes are rich in carotenoids and in vitamins C and B_6, while the apples add fiber.

- 4 large sweet potatoes
- 2 large, unpeeled apples
- 2 Tbsp (30 ml) vegan margarine, melted
- ¼ cup (60 ml) maple syrup mixed in ¼ cup (60 ml) hot water
- ½ tsp (3 ml) ground cloves
- 1 tsp (5 ml) cinnamon
- ½ cup (125 ml) apple juice
- 1 cup (250 ml) chopped pecans or walnuts (optional)

Instructions:

1. Preheat the oven to 350 °F (180 °C).
2. Cook the sweet potatoes in boiling water until done, but firm. Let cool, and peel.
3. Core and slice the apples to ¼-inch (6-mm) thickness.
4. Once the sweet potatoes have cooled sufficiently, cut into ½-inch (1.3 cm) slices.
5. Oil a deep, 1½ quart (1½ L) baking dish. Arrange half of the sweet potato slices on the bottom.
6. Drizzle with half of the margarine, then with half of the diluted maple syrup.
7. Top with half of the apple slices. Sprinkle lightly with half of each of the spices.
8. Repeat the layers as above, then pour the apple juice on top.
9. Bake for 30 minutes, covered, then for another 10 minutes, uncovered.
10. Sprinkle with nuts, if desired. Serve hot, or cover to keep warm.

Spinach and Mushroom Quiche (8 servings)

This vegan quiche has the texture and flavor of a traditional quiche, but without any of the saturated fat and cholesterol. As well, no manure was generated; water usage and greenhouse gas emissions were significantly reduced by not using dairy cheese.

14 oz (400 grams) extra-firm organic tofu
½ cup (125 ml) plant-based milk
1 Tbsp (15 ml) canola oil
1 small onion, diced
3 garlic cloves, minced
1 cup (250 ml) chopped fresh mushroom
2 cups (500 ml) chopped fresh spinach
½ tsp (3 ml) salt
½ tsp (3 ml) pepper
½ tsp (3 ml) turmeric
Sprinkle red pepper flakes
2 cups (500 ml) Daiya shreds

Instructions:

1. Preheat oven to 350 °F (180 °C).
2. In a food processor or blender, process the tofu and plant-based milk until creamy.
3. In a skillet, heat the oil and sauté the onion, garlic, and mushrooms for about 5 – 7 minutes.
4. Add spinach, salt, pepper, turmeric, and red pepper flakes. Cook until the liquid has evaporated (or drain off excess liquid).
5. Remove from heat and fold in Daiya shreds and the puréed tofu.
6. Pour into baking dish and bake for 40 minutes, or until the quiche is firm and has set.

"You have just dined, and however scrupulously the slaughterhouse is concealed in the graceful distance of miles, there is complicity."

~ Ralph Waldo Emerson

RESOURCES

Websites

American Institute for Cancer Research www.aicr.org
Michael Greger, MD www.nutritionfacts.org
Joel Fuhrman, MD www.drfuhrman.com
Vegan Nutrition, RD www.theveganrd.com
Jack Norris, RD www.veganhealth.org
Vegetarian Nutrition Dietetic Practice Group www.vegetariannutrition.net
Vegetarian Resource Group www.vrg.org
Physicians Committee for Responsible Medicine www.pcrm.org
Worldwatch Institute www.worldwatch.org
Environmental Working Group www.ewg.org
Mercy for Animals www.mercyforanimals.org
Plant-Built Vegan Muscle www.plantbuilt.com
Strongest Hearts web series on vegan athletes www.strongesthearts.org
Patricia Tallman, PhD www.drpatriciatallman.com
Matt Ruscigno, MPH, RD www.truelovehealth.com
Mark Rifkin, MS, RD www.balancednutritiononline.com

Report

Steinfeld, H.; Gerber, P.; Wassenaar, T.; Castel, V.; Rosales, M.; de Haan, C. *Livestock's Long Shadow: Environmental Issues and Options.* UN FAO: Rome, 2006

PDF: http://www.fao.org/docrep/010/a0701e/a0701e00.HTM

Books

Barnard, Neal. *The Reverse Diabetes Diet*. Emmaus, Pennsylvania: Rodale Books, 2010.

___________. *Power Foods for the Brain*. New York: Grand Central Publishing, 2013.

Davis, Brenda and Melina, Vesanto. *Becoming Vegan: Express Edition*. Summertown, Tennessee: Book Publishing Company, 2013.

Foer, Jonathan Safran. *Eating Animals*. Boston: Little, Brown and Company, 2009.

Keon, Joseph. *Whitewash – The Disturbing Truth about Cow's Milk and your Health.* Gabriola Island, British Columbia: New Society Publishers, 2010.

Lappé, Anna. *Diet For a Hot Planet: The Climate Crisis at the End of Your Fork and What You Can Do About It*. New York: Bloomsbury USA, 2011.

Messina, Virginia and Norris, Jack. *Vegan for Life*. Boston: Da Capo Press/Lifelong Books, 2011.

Moran, Victoria. *Main Street Vegan*. New York: Jeremy Tarcher/ Penguin, 2012.

______________. *The Good Karma Diet*. New York: Penguin Publishing Group, 2015.

Robbins, John. *No Happy Cows.* San Francisco: Conari Press, 2012.

Simon, David. *Meatonomic$: The Bizarre Economics of Meat and Dairy*. San Francisco: Conari Press, 2013.

Specific to vegan athletic performance

Frazier, Matt and Ruscigno, Matthew. *No Meat Athlete*. Beverly, MA: Fair winds Press, 2013.

Jurek, Scott. *Eat and Run.* New York, NY: Houghton Mifflin Harcourt, 2012.

Larson-Meyer, Enette. *Vegetarian Sports Nutrition.* Champaign, IL: Human Kinetics, 2007.

Moskowitz, Isa and Ruscigno, Matthew. *Appetite for Reduction*. Philadelphia, PA: Da Capo Press, 2011.

Roll, Rich. *Finding Ultra*. New York, NY: Crown Publishing Group, 2012.

Ruscigno, Mathew and Ploeg, Joshua. *Superfoods for Life, Cacao*. Beverly, MA: Fair Winds Press, 2014.

Documentary Films

***Cowspiracy: The Sustainability Secret* (2014)**

From the website (www.cowspiracy.com):

"*COWSPIRACY: The Sustainability Secret* is a ground-breaking, feature-length environmental documentary following an intrepid filmmaker as he uncovers the most destructive industry facing the planet today – and investigates why the world's leading environmental organizations are too afraid to talk about it."

***Earthlings* (2005)**

From the website (www.earthlings.com):

"*EARTHLINGS* is an award-winning documentary film about the suffering of animals for food, fashion, pets, entertainment and medical research. Considered the most persuasive documentary ever made, EARTHLINGS is nicknamed 'the Vegan maker' for its sensitive footage shot at animal shelters, pet stores, puppy mills, factory farms, slaughterhouses, the leather and fur trades, sporting events, circuses and research labs."

The film is narrated by Academy Award® nominee Joaquin Phoenix, and features music by platinum-selling recording artist Moby. Initially ignored by distributors, today *EARTHLINGS* is considered by organizations around the world to be the definitive animal rights film. "Of all the films I have ever made, this is the one that gets people talking the most," said Phoenix. "For every one person who sees *EARTHLINGS*, they will tell three."

APPENDIX

TABLE 1

Energy and Protein Conversion Efficiencies of Feed Inputs to Edible Meat in the US

(Excerpted from Table 4.1 in Smil, V. *Should We Eat Meat?* Chichester, Massachusetts: John Wiley & Sons, 2013. Vaclav Smil, Ph.D., a Professor Emeritus at University of Manitoba, was listed in 2010 by the journal *Foreign Policy* as one of the top 100 global thinkers.)

Measurement	Chicken	Pork	Beef
Feed (kg/kg Live Weight, LW)	2	5	10
Edible weight (% LW)	60	53	40
Feed (kg/kg Edible Weight, EW)	3.3	9.4	25
Energy content of feed (Mega Joules/kg)	15	15	15
Energy content of meat (MJ/kg EW)	7.5	13.0	13.4
Energy conversion efficiency (%)	15	9.2	3.6
Protein content of feed (%)	20	15	15
Protein content of meat (% of EW)	20	14	15
Protein conversion efficiency (%)	30	10	4
Feed energy per unit of protein (MJ/g)	2.5	10	25

DETERMINATION OF MANURE PER AMOUNT OF EDIBLE PRODUCT

Source: National Engineering Handbook, Part 651, Chapter 4, "Agricultural Waste Management Field Handbook," USDA Natural Resources Conservation Service (updated 2008). This was used for the following calculations.

Dairy Table 4-5 (a) (see Source)

The amount of manure produced by one cow over her lifetime (typically 5 years) is as follows:

a) 156 lb per day (average) x (2.5 years x 365) + 27 lb per day x (13* months x 30) + 54 lb per day (9 months x 30) + 85 lb per day (6 months during the 3 years x 30) = 142,350 + 10,530 + 14,580 + 15,300 = 182,760 lb of manure.
b) The average milk production is 6.5 gal per day. The number of gallons produced by the average cow is 6.5 gal x 2.5 years x 365 = 5,931 gallons.
c) Pounds of manure per gallon of milk = 182,760 ÷ 5,931 = 31 lb per gallon of milk.
d) 1 US gallon = 8.34 lb; thus, the amount of manure per pound of milk = 31 ÷ 8.34 = 3.7 lb of manure per lb of milk.
e) Since 10 lb of milk are required to make 1 lb of cheese; the manure per lb of cheese = 37 lb of manure per lb of cheese.

*During the first 6 – 8 weeks, calves are fed milk or a milk replacer.

Beef Tables 4-8 (a) and 4-8 (c) (see Source)

The amount of manure produced by one beef cow over a lifetime (typically 18 – 22 months) is as follows:

a) Assume 18 months for a minimum estimate. As shown in Table 4-8 (c), 9,800 lb of manure are produced during the last 5 months of the animal's life.
b) As shown in Table 4-8 (a), there is no data for manure production for the first 6 months (before weaning). From 6 month onwards, the growing calf produces 50 lb of manure per day. This growing (i.e., "feeder") stage is a period of (18 - 5 - 6) = 7 months. The manure produced over these 7 months, at 50 lb per day, is 10,500 lb.
c) Thus, a minimum estimate, using Table 4-8 data, is 9,800 + 10,500 = 20,300 lb of manure over the cow's lifetime.
d) The average weight of a beef cow is 1,300 lb. Thus, the amount of manure per edible pound of beef is at least = 20,300 ÷ (1.300 x 0.4 efficiency) = 39 lb of manure per lb of beef, or 39 kg of manure per kg of beef.

Swine **Table 4-10 (c)** (see Source)

The manure produced by one pig over its lifetime is as follows:

a) 87 lb of manure are produced during the nursery stage (at an average weight of 27.5. reaching 45 lbs) + 1,200 lb manure ("grow to finish" at an average weight of 154 lb, although the slaughter weight is 283 lb) = 1,287 lb.

b) Thus, the total manure output per edible weight of pork = 1,287 ÷ (283 x .53 efficiency) = 8.6 lb of manure per lb of pork, or 8.6 kg of manure per kg of pork.

Egg-Layers **Table 4-11 (a)** (see Source)

The manure produced by an egg-layer over its lifetime is as follows:

a) Pullets start producing eggs around 20 weeks (5 months). The average annual yield is 275 eggs per year (USDA, NASS), for approximately 12 – 14 months of egg-laying.

b) Table 4-11 (a) shows 0.19 lb of manure per day x (365 days) = 69 lb of manure in one year. The average egg production is 275 eggs per annum. Thus, the amount of manure per egg = 69 ÷ 275 = 0.25 lb per egg, or .11 kg per egg.

c) There are 9 Grade A Large eggs in 1 lb of eggs; 1 egg = 50 grams. Thus, the amount of manure per pound of eggs = .25 x 9 = 2.3 lb of manure per lb of eggs, or 2.3 kg of manure per kg of eggs.

Broilers **Table 4-11 (c)** (see Source)

The amount of manure produced by a broiler chicken over its lifetime (typically 5 – 6 weeks) is as follows:

a) 11 lb per finished animal. Average slaughter weight is 5.9 lb.

b) Thus, the amount of manure per pound of edible chicken meat = 11 ÷ (5.9 x 0.6 efficiency) = 3.1 lb of manure per lb of chicken meat, or 3.1 kg of manure per kg of chicken meat.

http://www.ansc.purdue.edu/faen/dairy%20facts.html (milk production per cow)
http://www.nass.usda.gov/Publications/Todays_Reports/reports/lyegan14.pdf (egg production per hen)

FOOTNOTES

Dry beans to cooked beans

1. 1 cup of dry beans is approximately 3 cups of cooked beans, weighing 8 oz (225 grams).

Soy Milk: GHG

1. GHG for SILK brand soy milk: 47% less than dairy = .53 of 1.9 (dairy) = **1 kg per liter**
2. GHG for Tesco brand of soy milk: 0.7 – 0.9 kg per liter.

Daiya vegan "cheese": GHG

1. Daiya's main ingredients: tapioca (from cassava), canola &/or safflower oil, coconut oil
2. From EWG: CO_{2e} for vegetables in general = 2 CO_{2e} per kg of food; Tapioca is made from cassava, a vegetable.
3. From *Shrink That Footprint*: 0.8 g CO_{2e} per cal of oil, spreads for canola/safflower and coconut oil => 0.8 x 9 cal per gram = 7.2 g CO_{2e} per gram = 7.2 kg CO_{2e} per kg [NB: the kcal notation is ⇔calories]; 2.8 g CO_{2e} per cal of vegetables, breads.
4. CO_{2e} for Daiya: per 28 grams, there are 7 g (carbs) + 6 g (oils/spreads) + 1 g (protein). => 7 (2.8x4) + 6 (0.8x9) + 1 (2.8x14.7)* = 78.4 + 43.2 + 41.2 = 162.8 g CO_{2e}. So for 1 kg, => (162.8x1000) ÷ 28 = 5,814 g or **5.8 kg of CO_{2e} per kg of Daiya shreds.**

Daiya vegan "cheese": Water Footprint

1. (coconut oil) + 1 g (pea protein) => 7 (2.818 liters per gram) + 4 (4.301 liters per gram) + 2 (4.49 liters per gram) + **1(.6 liters per gram) = 19.726 + 17.204 + 8.98 + .6 = 46.51 liters. Thus, for 1 kg of shreds => 46.51 (1000 ÷ 28) = **1,661 liters per kg of Daiya shreds**
2. *There are about 5.5 grams of protein in 100 grams of fresh or cooked peas => 5.5%
3. 100 grams of peas => 81 cal (USDA). Thus, 1 gram of pea protein => 100 ÷ 5.5 = 18.2 grams of peas => 81 ÷100 x 18.2 = 14.7 cal.
4. **For peas, the water footprint is 595 liters per kg, roughly .6 liters per gram. Pea protein is slightly more because of further processing, but is difficult to estimate. However, it's small in terms of the other components.

Eggs: Water Footprint

1. 1 lb eggs = .454 kg => 9.5 eggs; 0.05 kg per egg. Water footprint per egg = 3,265 liters per kg x .05 kg per egg = **163 liters per egg**.

"Flax Egg": GHG & Water Footprint

1. ***"Flax egg": 1 "Flax egg" uses 1 Tbsp ground flax seed + 3 Tbsp water
2. In the absence of carbon footprint for seeds, use the same as CO_{2e} for nuts: **2.3 kg CO_{2e} per kg**;
3. Water footprint for flax seed = linseed = 5,168 liters per kg; 1 Tbsp of ground flax seed ~ 7 grams = 0.007 kg. Thus, Water footprint for "flax egg" = (5,168 x .007) + .045 (from 3 Tbsp water) = **36.2 liters per "flax egg."**
4. 1 Tbsp flaxseed = 7 grams. Each "flax egg" = 45 grams water + 7 grams flaxseed = 52 grams. 1 kg of "flax egg" contains 865 grams of water + 135 grams of flaxseeds. 1 liter of water ~ 1 kg, .865 kg = .865 liters. Water footprint for "flax egg" = .865 + 5,168 (Water footprint for flaxseeds) x 0.135 = .865 + 698 = 699 liters per kg of "flax egg."

TABLE 2 Summary of Environmental Parameters

Per kg of food	*Kg of CO_2 equivalent	**Water footprint, Liters	***Manure production, kg
Lamb/ sheep/ goat	39.2	8,763	
Beef	27.0	15,415	39
Beef Burger (150 g)		2,350	
Soy burger (150 g)		158	0
Pork	12.1	5,988	8.6
Turkey	10.9	No data	
Chicken	6.9	4,325	3.1
Eggs	4.8	3,265 (163 liters per egg)	(Grade A large) 2.25 0.25 lb or .11 kg of manure / egg
"Flax egg" = 1 Tbsp flaxseed + 3 Tbsp water	Use the same as nuts in absence of data	698 + .865 = 699 (36 liters per flax egg)	0
Farmed salmon	11.9	---------	-------
Cheese	13.5	5,000	37
Daiya "cheese" shreds	5.8	1,661	0
Dairy	1.9	1,020	3.7
Soy milk	(SILK) 1.0	300	0
Tofu	2.0	572	0
Lentils	0.9	5,874	0
Beans (uncooked)	2.0	4,055	0
Starchy Roots	(potatoes) 2.9	387	0
Rice	2.7	(milled rice) 2,500	0
Cereals		1,644	0
Fruits		962	0
Broccoli	2.0	280	0
Nuts	2.3	9,063	0

*from EWG, Meat Eater's Guide to Climate Change and Health, 2011, Figure 1, US data

**from a) A.Y. Hoekstra, "The hidden water resource use behind meat and dairy" *Animal Frontiers*, April 2012, vol. 2, No. 2 (global averages)

**from b) M. M. Mekonnen & A. Y. Hoekstra, "The green, blue and grey water footprint of crop products" *Hydrology and Earth System Sciences*, 15, 1577-1600, 2011 (global averages)

**from c) some calculated as noted below in the footnotes (global averages)

***based upon Appendix, and converted to kg (USDA data).

REFERENCES

Chapter 1

1. Andrew Winston, "Local Food or Less Meat? Data Tells the Real Story," *Harvard Business Review Blog*, June 20, 2011.
2. US – EPA website on Climate Change: www.epa.gov/climatechange/science/future.html
3. Gainesville.com, "Studies: Wildfires worse due to global warming," *The Associated Press*, May 18, 2014.
4. UN - FAO, "Livestock's Long Shadow – Executive Summary," Chapter 4: "Livestock's role in water depletion and pollution," November 2006.
5. Kari Hamerschlag, "Meat Eater's Guide to Climate Change + Health," Environmental Working Group, July 2011.
6. Laura Reynolds, "Agriculture and Livestock Remain Major Sources of Greenhouse Gas Emissions," *Worldwatch Institute*, May 8, 2013.
7. EPA, "Greenhouse Gas Emissions from a Typical Passenger Vehicle," December 2011.
8. UNEP – Global Environmental Alert Service, "Growing Greenhouse Gas Emissions Due to Meat Production," October 2012.
9. UNEP, "Growing Greenhouse Gas Emissions Due to Meat Production," October 2012.
10. Tom Levitt, "What is the environmental footprint of mega-dairy farming?" *The Ecologist*, November 15, 2010.
11. Lindsay Wilson, "The carbon footprint of 5 diets compared," Shrink That Footprint.com, (accessed February 2014).
12. EPA, Greenhouse Gas Equivalencies Calculator, http://www.epa.gov/cleanenergy/energy-resources/calculator.html
13. ThinkGlobalGreen.org (accessed March 2013).
14. Gidon Eshel, "Grass-fed beef packs a punch to environment," *Reuters* blog, April 8, 2010.
15. Shrinkthatfootprint.com, "The Truth about Food miles," (accessed January 2014).
16. Christopher L. Weber and H. Scott Matthews, "Food-Miles and the Relative Climate Impacts of Food Choices in the United States," *Environmental Science & Technology*, 2008: 42 (10), 3508–3513.
17. Katie Valentine, "Not Eating Meat Can Cut Your Food-Related Carbon Emissions Almost in Half, Study Finds," Climate Progress, http://thinkprogress.org/climate/2014/06/27/3454129/eating-meat-carbon-emissions/ June 27, 2014.
18. Nathan Fiala, "How Meat Contributes to Global Warming," *Scientific American*, February 2009.

19. Peter Scarborough, Paul N. Appleby, et. al., "Dietary greenhouse gas emissions of meat-eaters, fish-eaters, vegetarians, and vegans in the UK," *Climatic Change*, published online June 11, 2014.
20. Gidon Eshel, et. al.,"Land, irrigation water, greenhouse gas, and reactive nitrogen burdens of meat, eggs, and dairy production in the United States," *Proceedings of the National Academy of Sciences,* Abstract, June 23, 2104.
21. Ed Bedington, "US beef industry lambasts environment link study," July 23, 2014, www.globalmeatnews.com
22. Ertug Ercin, Maite M. Aldaya, Arjen Y. Hoekstra, "The water footprint of soy milk and soy burger and equivalent animal products," *Ecological Indicators,* 18 (2012): 392 – 402.
23. Arjen Y. Hoekstra, "The hidden water resource use behind meat and dairy," *Animal Frontiers*, April 2012, vol. 2, no. 2.
24. M. M. Mekonnen, A.Y. Hoekstra, "The green, blue and grey water footprint of crops and derived crop products," *Hydrology and Earth System Sciences*, 15, 1577-1600, 2011.
25. Vaclav Smil, *Should We Eat Meat?* John Wiley & Sons, 2013.
26. Waterfootprint.org, "Product Water Footprints" page (accessed January 2014).
27. FEW Resources.org, Water Scarcity Issues (accessed December 2013).
28. Richard Oppenlander, "Animal agriculture: A huge waste of water," thescavenger.net, October 23, 2011.
29. Fiona Harvey, "Eat less meat for greater food security, British population urged," *The Guardian*, June 4, 2013.
30. Mark Prigg, "Food shortages could turn most of the world vegetarian by 2050, warn leading scientists," *The Daily Mail*, August, 27, 2012.
31. John Vidal, "Food shortages could force world into vegetarianism, warn scientists," *The Guardian*, August 26, 2012.
32. *World Watch Magazine*, "Is Meat Sustainable?" by the Editors, July/August 2004, Vol. 17, No. 4.
33. Jason Sarasota, "How Does Meat in the Diet Take an Environmental Toll?" *Scientific American*, December 28, 2011.
34. Felicity Carus, "UN urges global move to meat and dairy-free diet," *The Guardian*, June 2, 2010.
35. Mark Bittman, "Rethinking the Meat-Guzzler," *New York Times*, January 27, 2008.
36. Huffingtonpost.com, "Vegetarian Diet Needed to Prevent Global Food and Water Crisis, Report Says," August 28, 2012.
37. Diets in Review.com, "Vegetarian Diet May Be Necessary to Prevent Global Water and Food Shortage," August 29, 2012.
38. Alastair Bland, "Is the Livestock Industry Destroying the Planet?" Smithsonian.com, August 1, 2012.
39. L.V. Anderson, "What would happen if everyone in the world gave up meat?" *Business Insider*, May 3, 2014.

40. Worldhunger.org, "2012 World Hunger and Poverty Facts and Statistics."
41. Will Dunham, "Weight of the World: 2.1 billion people obese or overweight,"*Reuters US*, May 28, 2014.
42. USDA, Natural Resources Conservation Service, "Animal Manure Management."
43. *National Engineering Handbook*, Part 651, Chapter 4, "Agricultural Waste Management Field Handbook," USDA Natural Resources Conservation Service, (updated 2008).
44. Leslie Kaufman, "Chemicals in Farm Runoff Rattle States on the Mississippi," *New York Times*, June 2, 2011.
45. David Biello, "Oceanic Dead Zones Continue to Spread," *Scientific American*, August 15, 2008.
46. FoodPolitics.com (accessed January 2014).
47. Natural Resources Defense Council (NRDC) "Reform Wildlife Services' Predator Control" last update 2/11/2013, (accessed January 2014), www.nrdc.org
48. Roddy Scheer and Doug Moss, "The USDA's 'Predator Control' Program", *E – the Environmental Magazine* www.emagazine.com November 18, 2012.
49. Stephanie Feldstein, "Hey Wildlife Lovers: Take Extinction Off Your Plate," *The Huffington Post*, March 20, 2014.
50. Paula Crossfield, "A New Report Reveals that GM Seeds Encourage Pesticides Use, Contribute to Growth of Superweeds," Civileat.com, November 17, 2011.
51. CnAgri (*China Agricultural Report*), "UN study reveals increase in herbicide use on GM crops," October 8, 2012.
52. Food & Water Europe: press release, "EU Must Draw a Line Under GMOs as Superweeds, Herbicide Use Soar," July 9, 2013.
53. Susan Mann, "Growth promotant claims to be removed from antimicrobial products," *Better Farming*, April 17, 2014.
54. Natural Resources Defense Council, "Saving Antibiotics"(accessed April 2014), www.nrdc.org
55. Margaret Munro, "Canada's chicken farmers ban infections that trigger superbugs," Canada.com, April 17, 2014.

Chapter 2 (contributed by Matt Ruscigno, MPH, RD)

1. N.D. Barnard, A. Nicholson, J.L. Howard JL, "The medical costs attributable to meat consumption. Preventive Medicine," 24(6): 646-55, Nov 1995.
2. P.W. Sullivan, V. Ghushchyan, et. al., "The medical cost of cardiometabolic risk factor clusters in the United States," *Obesity* (Silver Spring), 15(12): 3150-8, Dec 2007.

3. P. Anand, A.B. Kunnumakkara, et. al., "Cancer is a preventable disease that requires major lifestyle changes," *Pharmaceutical Research*, 25(9): 2097-116, Sep 2008.
4. J.F. Gonzales, N.D. Barnard, et. al., "Applying the Precautionary Principle to Nutrition and Cancer," *Journal of the American College of Nutrition*, 33:3, 239-246, 2014.
5. N.D. Barnard, H.I. Katcher, et. al., "Vegetarian and vegan diets in type 2 diabetes management," *Nutrition Review*, 67(5): 255-63, May 2009.
6. H. Kahleova, M. Matoulek, et. al., "Vegetarian diet improves insulin resistance and oxidative stress markers more than conventional diet in subjects with Type 2 diabetes," *Diabetic Medicine*, 28(5): 549-59, May 2011.
7. P.V. Babu, D. Liu, E.R. Gilbert, "Recent advances in understanding the anti-diabetic actions of dietary flavonoids," *Journal of Nutritional Biochemistry*, S0955-2863(13) 00127-7, Sep 9, 2013.
8. C.B. Esselstyn, G. Gendy, et. al., "A way to reverse CAD?" *Journal of Family Practice*, 63(7):56-364b, July 2014.
9. Z. Chen, PP. Wang, et. al., "Dietary patterns and colorectal cancer: results from a Canadian population-based study," *Nutrition Journal*, 14:8, January 15, 2015.
10. H. Iwase, M. Tanaka, et. al., "Lower vegetable protein intake and higher dietary acid load associated with lower carbohydrate intake are risk factors for metabolic syndrome in patients with type 2 diabetes: Post hoc analysis of a cross-sectional study," *Journal of Diabetes Investigation*, E-published January 4, 2015.
11. M.J. Orlich, P.N. Singh, et. al., "Vegetarian dietary patterns and mortality in Adventist Health Study 2," *JAMA Internal Medicine*, 173(13): 1230-8, Jul 8, 2013.
12. M. Sagner, D. Katz, et. al., "Lifestyle medicine potential for reversing a world of chronic disease epidemics: from cell to community," *International Journal of Clinical Practice*, 68(11): 1289-92, Nov 2014.
13. N. Wilson, N. Nghiem, "Foods and dietary patterns that are healthy, low-cost, and environmentally sustainable: a case study of optimization modeling for New Zealand," *PLoS One*, 8(3): e59648, 2013.
14. D. Mudgil, S. Barak, "Composition, properties and health benefits of indigestible carbohydrate polymers as dietary fiber: A review," *International Journal of Biological Macromolecules* 61C: 1-6, Jul 2, 2013.

15. D. Aune, D.S. Chan, et. al., “Dietary fibre, whole grains, and risk of colorectal cancer: systematic review and dose-response meta-analysis of prospective studies,” *British Medical Journal*, 343: d6617, Nov 10, 2011.
16. L. Schwingshackl, B. Strasser, et. al., “Effects of monounsaturated fatty acids on cardiovascular risk factors: a systematic review and meta-analysis,” *Annals of Nutrition Metabolism*, 59(2-4): 176-86, 2011.
17. P.W. Siri-Tarino, Q. Sun Q, et. al., “Saturated fatty acids and risk of coronary heart disease: modulation by replacement nutrients,” *Current Atherosclerosis Reports*, 12(6): 384-90, Nov 2010.
18. G.D. Lawrence, “Dietary fats and health: dietary recommendations in the context of scientific evidence,” *Advances in Nutrition*, 4(3): 294-302, May 1, 2013.
19. V. Habauzit, C. Morand, et. al., “Evidence for a protective effect of polyphenols-containing foods on cardiovascular health: an update for clinicians,” *Therapeutic Advances in Chronic Disease*, 3(2): 87-106, Mar 2012.
20. Y. Song, N.R. Cook, C.M. Albert, et. al., “Effects of vitamins C and E and beta-carotene on the risk of type 2 diabetes in women at high risk of cardiovascular disease: a randomized controlled trial,” *American Journal of Clinical Nutrition*, 90(2): 429–437, 2009.
21. Y. Ye, J. Li, Z. Yuan, “Effect of antioxidant vitamin supplementation on cardiovascular outcomes: a meta-analysis of randomized controlled trials,” *PLoS One*, 8(2): e56803, E-published Feb 20, 2013.
22. J. Fuhrman, D.M. Ferrerl, “Fueling the vegetarian (vegan) athlete,” *Current Sports Medicine* - Reports., Vol. 9, No. 4, pp. 233-241, 2010.
23. L. Fontana, T. E. Meyer, S. Klein, J.O. Holloszy, “Long-term low-calorie low-protein vegan diet and endurance exercise are associated with low cardiometabolic risk,” *Rejuvenation Research*, 10(2): 225-34, Jun 2007.
24. American College of Sports Medicine, American Dietetic Association and Dietitians of Canada. (2009) “Nutrition and athletic performance. Joint position statement,” *Medicine and Science in Sports and Exercise* 41, 709-731, March 2009.
25. B. Bolling, D.L. McKay, J.B. Blumberg, “The phytochemical composition and antioxidant actions of tree nuts,” *Asia Pacific Journal of Clinical Nutrition*, 19 (1): 117-123, 2010.
26. P. Kris-Etherton, “Walnuts decrease risk of cardiovascular disease: a summary of efficacy and biologic mechanisms,” *Journal of Nutrition*, Apr 2014, 144(4 Suppl): 547S-554S, E-published Feb 5, 2014.
27. E. Ros, “Nuts and novel biomarkers of cardiovascular disease,” *American Journal of Clinical Nutrition*, May 2009, 89(5): 1649S-56S, E-published Mar 25, 2009.

28. I.Fistonić I, M. Situm, et. al. "Olive oil biophenols and women's health," *Medicinski glasnik (Zenica),* 9(1): 1-9. Feb 2012.
29. S. Silva, M.R. Bronze, et. al., "Impact of a 6-wk olive oil supplementation in healthy adults on urinary proteomic biomarkers of coronary artery disease, chronic kidney disease, and diabetes (types 1 and 2): a randomized, parallel, controlled, double-blind study," *American Journal of Clinical Nutrition*, Jan 2015, 101(1): 44-54, E-published Nov 19, 2014.
30. E. Waterman, B. Lockwood, "Active components and clinical applications of olive oil," *Alternative Medicine Review*, Dec 2007 Dec, 12(4): 331-42.
31. University of Maryland, Medical Center, webpage, accessed January 24, 2015, http://umm.edu/health/medical/altmed/supplement/soy
32. L. Yan, E.L. Spitznagel, "Soy consumption and prostate cancer risk in men: a revisit of a meta-analysis," *American Journal of Clinical Nutrition*, 89(4): 1155-63, Apr 2009.
33. X.F. Zhao, L.Y. Hao, et. al., "A study on absorption and utilization of calcium, iron and zinc in mineral-fortified and dephytinized soy milk powder consumed by boys aged 12 to 14 years," *Zhonghua Yu Fang Yi Xue Za Zhi (in Chinese)*, 37(1): 5-8, Jan 2003.
34. M. Messina, C. Nagata, A.H. Wu, "Estimated Asian adult soy protein and isoflavone intakes," *Nutrition and Cancer*, 55(1): 1-12, 2006.
35. L.E. Murray-Kolb, R. Welch, et. al., "Women with low iron stores absorb iron from soybeans," *American Journal of Clinical Nutrition*, 77(1): 180-4, Jan 2003.
36. L.K. Beaton, B.L. McVeigh, et. al., "Soy protein isolates of varying isoflavone content do not adversely affect semen quality in healthy young men," *Fertility and Sterility*, 94(5): 1717-22, Oct 2010.
37. M. Messina, S. Watanabe, K.D. Setchell, "Report on the 8th International Symposium on the Role of Soy in Health Promotion and Chronic Disease Prevention and Treatment," *Journal of Nutrition*, 139(4): 796S-802S, Apr 2009.
38. H.W. Lopez, F. Leenhardt, et. al.,"Minerals and phytic acid interactions: is it a real problem for human nutrition?" *International Journal of Food Science & Technology*, 37: 727–739, 2002.
39. A. Ørgaard, L. Jensen, "The Effects of Soy Isoflavones on Obesity," *Exp. Biol. Med*, 233: 1066-1080, 2008.
40. Food and Drug Administration, Food Labeling: Health Claims; Soy Protein and Coronary Heart Disease. October 26, 1999.
41. J.M. Hamilton-Reeves, G. Vazquez, et. al., "Clinical studies show no effects of soy protein or isoflavones on reproductive hormones in men: results of a meta-analysis," *Fertility and Sterility*, Jun 11, 2009.
42. L.A. Korde, A.H. Wu, et. al., "Childhood Soy Intake and Breast Cancer Risk in Asian American Women," *Cancer Epidemiology, Biomarkers & Prevention*, 18(4); 1050-9, April 2009.
43. J.E. Chavarro, T.L. Toth, S.M. Sadio, R. Hauser, "Soy food and isoflavone intake in relation to semen quality parameters among men from an

infertility clinic," *Human Reproduction,* 23 (11): 2584–90, November 2008.

44. S.J. Nechuta, B.J. Caan, W.Y. Chen, et. al., "Soy food intake after diagnosis of breast cancer and survival: an in-depth analysis of combined evidence from cohort studies of US and Chinese women," *American Journal of Clinical Nutrition*., 96(1): 123-132, 2012.
45. T. Oseni, et. al., "Selective Estrogen Receptor Modulators and Phytoestrogens," *Planta Medica*, 74(13): 1656-1655, 2008.

Chapters 3 – 6

1. EPA, Greenhouse Gas Equivalencies Calculator, http://www.epa.gov/cleanenergy/energy-resources/calculator.html
2. USDA National Agricultural Statistics Service, "Farm Animal Statistics: Slaughter Totals," for years up to 2013, April 17, 2014.
3. Gidon Eshel, et. al., "Land, irrigation water, greenhouse gas, and reactive nitrogen burdens of meat, eggs, and dairy production in the United States," *Proceedings of the National Academy of Sciences,* Abstract, June 23, 2104.
4. M.L. McCullough, S.M. Gapstur, R. Shah, E.J. Jacobs, P.I. Campbell, "Association between red and processed meat intake and mortality among colorectal cancer survivors," *J. Clinical Oncology*, published online July 1, 2013.
5. M.S. Farvid, E. Cho, W.Y. Chen, H.H. Eliassen, W.C. Willet, "Dietary protein sources in early adulthood and breast cancer incidence: prospective cohort study," *British Medical Journal*, published online June 10, 2014.
6. Pan, Q. Sun, A.M. Berstein, J.E. Manson, W.C. Willett, F.B. Hu, "Changes in red meat consumption and subsequent risk of type 2 diabetes mellitus," *JAMA Intern Med*. published online, June 17, 2013.
7. J. Kaluza, A. Akeson, A. Wolk, "Processed and unprocessed red meat consumption and risk of heart failure: a prospective study of men," *Circulation Heart Failure*, published on-line June 12, 2014.
8. Canada Beef Inc., Corporate Info, 2012.
9. Northwest Farm Credit Services, "Industry Perspective – FEEDLOT," prepared by the Northwest FCS Cattle Knowledge Team, 2007.
10. Natural Resources Defense Council, http://www.nrdc.org/water/pollution/ffarms.asp (last updated Feb 21, 2013).
11. Andy Wright, "Pigheaded: How Smart are Swine?" *Modern Farmer*, March 10, 2014.

12. PEW Environment Group, Joshua S. Reichert, Managing Director "Big Chicken: Pollution and Industrial Poultry Production in America," July 26, 2011.
13. Nicholas Bakalar, "Risks: More Red Meat, More Mortality," *New York Times*, March 12, 2012.
14. www.NutritionFacts.org video concerning Turmeric and Pancreatic Cancer, October 24, 2014.
15. US National Institutes of Health, National Cancer Institute, NIH – AARP Diet & Health Study,www.cancer.gov
16. Terrence O'Keefe, "Look for a big jump in broiler weights this year," WATTAgNet.com, Food Safety and Processing Perspective, May 15, 2013.
17. Stephane Perrais, "Franken Chickens Grow 6 times faster than 100 years ago," Mercy For Animals Canada blog, September 17, 2013.
18. HSUS "Farm Animal Statistics: Slaughter Total,"July 3, 2014.
19. R.A. Koeth, Z.Wang, B.S. Levison, et.al., "Intestinal microbiota metabolism of L-carnitine, a nutrient in red meat, promotes atherosclerosis," *Nat. Med.* published online April 7, 2013.
20. W.H.W. Tang, Z. Wang, B.S. Levison, et.al., "Intestinal microbial metabolism of phosphatidylcholine and cardiovascular risk," *N. Engl. J. Med.* 2013; 368: 1575-1584.
21. Karen Davis, Ph.D., "Are Chickens Smarter than Toddlers?" *United Poultry Concerns*, August 19, 2013.
22. Colin Smith, original article source, "Bird brain? Birds and humans have similar brain wiring," *Science Daily*, July 17, 2013 (based on materials provided by Imperial College London).
23. Fiona Macrae, "Can chickens REALLY be cleverer than a toddler? Studies suggest animals can master numeracy and basic engineering," Daily Mail Online, September 23, 2014.
24. Rachel Payne, "Chickens and other birds are smarter than you think," examiner.com, May 14, 2012.
25. Katherine Martinko, "Chickens out-perform toddlers in math tests," TreeHugger.com, October 21, 2013.
26. Joseph Keon, *Whitewash – The Disturbing Truth about Cow's Milk and Your Health*, New Society Publishers, 2010.
27. John Robbins, *No Happy Cows*, Conari Press, 2012.
28. Daniel More, M.D. "Top 7 food Allergies in Children," About.com - Allergies, Updated June 15, 2007.
29. http://www.foodreactions.org/intolerance/lactose/prevalence.html "Prevalence, Age, Gender & Genetics," accessed June 2012.
30. Elizabeth Weise, "Sixty percent of adults can't digest milk," *USA Today*, Sept 15, 2009.
31. AllergicLiving.com, "The Prevalence of Milk and Egg Allergies," accessed September 2014, http://allergicliving.com/2010/08/24/statistics-milk-allergy-eggs-allergy/

32. Neal Barnard, *Breaking the Food Seduction*, New York: St. Martin's Press, 2003.
33. Mark Astley, "Greek yogurt waste 'acid whey' a concern for USDA," Jones Laffin, Dairyreporter.com, Jan 30, 2013.
34. Rachel Cernansky, "What does our taste for Greek Yogurt mean for the Environment?" Planet Green, http://recipes.howstuffworks.com/what-does-our-taste-for-greek-yogurt-mean-for-the-environment.htm
35. Justin Elliott, "Whey too much: Greek Yogurt's Dark Side," Modern Farmer, modernfarmer.com, May 22, 2013.
36. Julia Lurie and Alex Park, "It takes HOW much water to make Greek Yogurt?!" Mother Jones, motherjones.com, Mar 10, 2014.
37. Cherrill Hicks, "Cows' milk is good for calves, but not for us: Avoid cancer by axing dairy, meat products, U.K. scientist and six-time survivor urges," *The Daily Telegraph*, June 2, 2014.
38. Meredith Engel, "New nutrition guidelines recommend foods for preventing cancer," *New York Daily News*, June 10, 2014.
39. Sophie Borland, "Breast cancer patients who eat cheese, yogurts or ice cream could HALVE their chances of survival," *Daily Mail*, March 14, 2013.
40. Y. Li, C. Zhou, X. Zhou, L. Li, "Egg consumption and risk of cardiovascular diseases and diabetes: A meta-analysis," *Atherosclerosis*, published ahead of print April 17, 2013.
41. Stephanie Brown and John Youngman, "The disgraceful secret down on the farm," *Ottawa Citizen*, October 10, 2003.
42. Louise Gray, "40 million chicks on 'conveyor belt to death,"*The Telegraph*, November 4, 2010.
43. RSPCA, "What happens with male chicks in the egg industry?" accessed August, 2014, http://kb.rspca.org.au/What-happens-with-male-chicks-in-the-egg-industry_100.html
44. United Poultry Concerns, "The Male Chick of the Egg Industry: He is Treated Like Trash," accessed August, 2014.
45. Syd Baumel, "Indecent Eggsposure: How Eggs are Laid in Canada," United Poultry Concerns, June 1, 2006.
46. Farm Sanctuary, Chickens used for Eggs, accessed August, 2014. http://www.farmsanctuary.org/learn/factory-farming/chickens/
47. Jonathan Balcombe, Ph.D. "The Inner Lives of Animals," *Psychology Today*, August 18, 2010.
48. *Mailonline*, "Moo-dini: Cow with 'unusual intelligence' opens farm gate with tongue so herd can escape shed," June 22, 2011.
49. Jerry Schleicher, "Farm Animal Intelligence: How Smart are your Cows?" GRIT, Rural American Know-How, November 2011. http://www.grit.com/animals/farm-animal-intelligence.aspx
50. Amy Armstrong, "How Smart are Cattle?" Animals by Demand Media, http://animals.pawnation.com/smart-cattle-3300.html

Chapter 7

1. Milton Mills, M.D., "The Comparative Anatomy of Eating," Vegsource.com, http://www.vegsource.com/news/2009/11/the-comparative-anatomy-of-eating.html , November, 21, 2009.
2. WHO Technical Report Series - 935, "Protein and Amino Acid Requirements in Human Nutrition," 2007.
3. Rob Dunn, "Human Ancestors Were Nearly All Vegetarians," *Scientific American (blog)*, July 23, 2012.
4. Gemma Tarlach, "Stone Age Farmers Showed Sophisticated Use of Fertilizers," discovermagazine.com, July 15, 2013, http://blogs.discovermagazine.com/d-brief/2013/07/15/stone-age-farmers-showed-sophisticated-use-of-fertilizers/#.VJScRl4AKA
5. Jessica Jones, MS, RD, "The Protein Myth: Why you need less protein than you think," Huffingtonpost.com, HuffPostHealthy Living, September 20, 2012.
6. Nation Center for Health Statistics – Blog, "Adults' daily protein intake much more than recommended," March 3, 2010.
7. John McDougall, M.D. "The Paleo Diet is Uncivilized (And Unhealthy and Untrue),"*The McDougall Newsletter*, June 2012.
8. Mary Beth O'Leary, "Controlling protein intake may be key to longevity, studies show," Elsevier.com, elsevierconnect, March 4, 2014.
9. Levine et. al., "Low Protein Intake is Associated with a Major Reduction in IGF-1, Cancer, and Overall Mortality in the 65 and Younger but Not Older Population," *Cell Metabolism*, March 2014.
10. Solon-Bietet, et. al., "The ratio of macronutrients, not caloric intake, dictates cardiometabolic health, aging and longevity in ad libitum-fed mice," *Cell Metabolism*, March 2014.
11. Sarah Knapton, "High-protein diet 'as bad for health as smoking," *The Telegraph*, March 4, 2014.
12. Anna Pippus, "Doctors Prescribe Veganism as 'Solution' to Heart Disease," Mercy For Animals Canada blog, May 23, 2014.
13. Sarah Knapton, "High protein diet linked to spiked cancer risk akin to smoking 20 cigarettes a day: U.S. Study," *The Daily Telegraph*, March 5, 2014.

Appendix

1. US EPA, Ag101, Life Cycle of a Cow, last updated June 27, 2012, http://www.epa.gov/oecaagct/ag101/dairyphases.html
2. Purdue University, Purdue Agriculture, Food Animal Education Network, "Dairy Facts" (accessed May 2014, http://www.ansc.purdue.edu/faen/dairy%20facts.html

3. BC Agriculture, "Eggs," accessed June 2014, http://www.agf.gov.bc.ca/aboutind/products/livestck/eggs.htm
4. Government of Saskatchewan, Agriculture, "Introduction to Poultry Production in Saskatchewan," http://www.agriculture.gov.sk.ca/Introduction_Poultry_Production_Saskatchewan accessed June 2013
5. USDA, "Livestock Slaughter," released by National Agricultural Statistics Services (NASS), Agriculture Statistics Board, USDA, April 24, 2014.
6. National Engineering Handbook, Part 651, Chapter 4, "Agricultural Waste Management Field Handbook," USDA Natural Resources Conservation Service, (updated 2008).

50918050R00112

Made in the USA
Charleston, SC
11 January 2016